飞行器设计与工程力学品牌专业系列教材

航空科技英语

Technical English for Aeronautics

主　编　沈　星

副主编　古　彪　许明轩

科　学　出　版　社

北　京

内 容 简 介

本书是为航空航天类专业的学生编写的双语教材，不仅为学生提供基本的航空知识和理论，更能帮助学生了解基本国际航空技术及相关的英语知识。

本书从航空简史出发，对航空技术做了较为系统的介绍。包括飞机的结构类型、组成及材料，飞机飞行的相关原理和控制技术，飞机的动力装置、进气系统、燃油系统、起动系统、润滑系统、制冷系统、排气系统、电力系统、液压系统、供氧系统、防冰装置，飞机内部的仪表装置，最后两章作为特种飞行器的专题部分，对除普通飞机外的特殊飞行器进行了科普性的说明，拓宽读者的航空知识。本书每章都采取中英文结合的形式，供读者查阅。

本书可作为高等学校飞行器设计与工程、飞行器环境与生命保障工程、工程力学等专业的教材，也可供航空航天爱好者参考与学习。

图书在版编目(CIP)数据

航空科技英语 = Technical English for Aeronautics / 沈星主编. —北京：科学出版社, 2020.3

飞行器设计与工程力学品牌专业系列教材

ISBN 978-7-03-061116-1

Ⅰ. ①航… Ⅱ. ①沈… Ⅲ. ①航空-科学技术-英语-高等学校-教材 Ⅳ. ①V2

中国版本图书馆 CIP 数据核字(2019)第 080972 号

责任编辑：余 江 张丽花 / 责任校对：彭珍珍

责任印制：吴兆东 / 封面设计：迷底书装

科学出版社 出版

北京东黄城根北街 16 号

邮政编码：100717

http://www.sciencep.com

北京凌奇印刷有限责任公司 印刷

科学出版社发行 各地新华书店经销

*

2020 年 3 月第 一 版 开本：787×1092 1/16

2023 年12月第四次印刷 印张：15

字数：350 000

定价：69.00 元

(如有印装质量问题，我社负责调换)

飞行器设计与工程力学品牌专业系列教材

编委会

丛 书 序

飞行器是现代最快速的交通工具，是现代战争重要的空中平台和武器装备，是人类探索宇宙的重要工具，因此，飞行器在军民两用和人类发展中都具有十分重要的地位，飞行器技术已成为现代高科技的重要标志。飞行器的核心技术是飞行器设计，随着科学技术的不断发展和对飞行器需求的不断增加，飞行器设计呈现出了快速发展的趋势，同时面临着许多挑战。飞行器设计的基础是工程力学，以飞行器为背景的工程力学伴随着飞行器设计技术的发展而发展，两者相辅相成，互相促进，共同发展。

南京航空航天大学的飞行器设计与工程、工程力学两个本科专业，以航空宇航科学与技术、力学两个一级学科国家重点学科为依托，以航空航天事业的建设者和开拓者为人才培养目标，持续不断地进行教育教学改革，为国家培养了一大批飞行器设计创新人才，校友中涌现出许多国家重点飞行器型号的设计师，在校学生也屡获国内外大学生创新竞赛的冠军。

近年来，随着教育教学改革的不断深入，学校进一步着重建设基础、创新、实践、国际“四位一体”的人才培养体系，加快推进专业建设和人才培养。2015 年，江苏省启动了“江苏高校品牌专业建设工程一期项目”，南京航空航天大学的飞行器设计与工程、工程力学两个专业双双入选 A 类，又为这两个专业建设和人才培养注入了新动力，乘此契机，编写出版“飞行器设计与工程力学品牌专业系列教材”，具有十分重要的意义。

本系列教材突出飞行器设计的专业基础和专业知识的系统性，突出飞行器设计的传统经典理论与现代技术方法的结合，体现南京航空航天大学的飞行器设计与工程、工程力学两个专业“围绕航空航天高科技、注重理工融合、突出工程实践”的特色。

希望本系列教材能成为我国飞行器设计的特色教材，能为我国飞行器设计创新人才的培养做出一定贡献。同时，希望大家对本系列教材提出宝贵意见，使之更加完善。

编委会

2016 年 8 月

前　言

以航空航天类专业为主的院校，其学生应具备基本的航空知识和理论，这是基本的专业要求。随着经济全球化的进一步发展，社会对当代大学生的英语水平也提出了更高的要求。编者希望这本《航空科技英语》教材，能够帮助学生了解基本的国际航空技术，掌握航空方面的专业英语知识和词汇，培养独立学习和研究的基本技能，达到提高专业水平的目的。同时，也希望通过学习本书，学生能阅读相关英文文献，为学习国内外先进技术或者出国深造打下基础。

本书以《航空航天概论》、《世界飞机手册》、*The Pilot's Handbook of Aeronautical Knowledge* 等中英文资料为基础，结合当前大学本科阶段学生的英语水平，从航空简史出发，对航空技术做了较为系统性的介绍。本书共 7 章，第 1～5 章为飞机基础知识介绍，分别就飞机结构、飞行知识、飞机系统以及飞行仪表等进行叙述；第 6、7 章为特种飞行器的专题介绍，对除普通飞机外的特殊飞行器进行了科普性的说明。本书旨在引入新的学科体系，融合中西文化，故每章都采取中英文结合的形式，配以大量图片资料来增强可读性，部分章节单独设有关键词汇和短语选解。

感谢为本书出版做出贡献的相关人员，特别是南京航空航天大学的金海波副教授、李杰锋副研究员、王韬熹讲师对本书的初稿提出了很多宝贵的意见，夏雪、王鹏军、逢宇翔、杨洋、黄赟同学参与了全文的校对工作。航空工业特种飞行器研究所的何卫平高级工程师等提供了很多相关参考资料，在此表示衷心的感谢。

限于编者水平，书中难免存在不妥之处，望读者指正，提出宝贵意见。

编者

2019 年 5 月

目　　录

Chapter 1 Introduction

1.1 History of Aviation

The identity of the first "birdmen" who fitted themselves with wings and leapt off a cliff in an effort to fly is lost in time. But each failure provided a reference for those who wished to fly. Where had the wing flappers gone wrong? Philosophers, scientists, and inventors offered solutions, but no one could add wings to the human body and soar like a bird. During the 1500s, Leonardo da Vinci filled pages of his notebooks with sketches of proposed flying machines, but most of his ideas were flawed because he clung to the idea of birdlike wings (Figure 1-1). By 1655, mathematician, physicist, and inventor Robert Hooke concluded that the human body does not possess the strength to power artificial wings. He believed human flight would require some form of artificial propulsion.

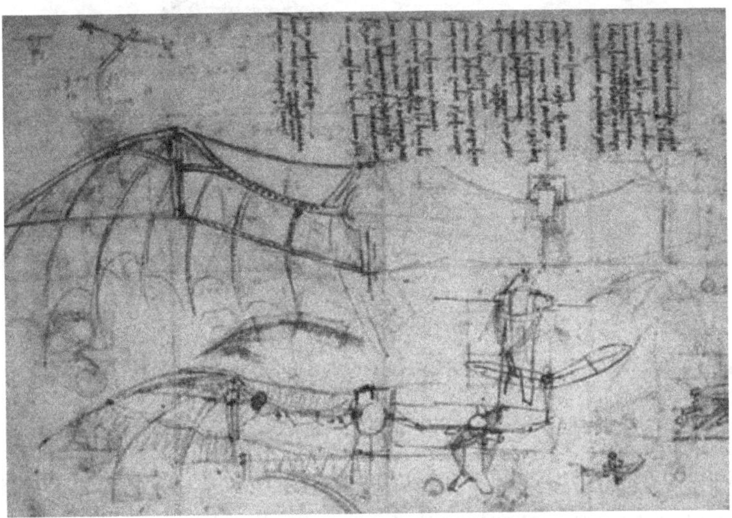

Figure 1-1 Leonardo da Vinci's ornithopter wings

The quest for human flight led some practitioners in another direction. In 1783, the first manned hot air balloon, crafted by Joseph and Etienne Montgolfier flew for 23 minutes. Ten days later, Professor Jacques Charles flew the first hydrogen balloon. Madness for balloon flight captivated the public's imagination and for a time flying enthusiasts turned their expertise to the promise of lighter-than-air flight.

Balloons solved the problem of lift, but that was only one of the problems of human flight. The ability to control speed and direction eluded balloonists. The solution to that problem lay in a child's toy familiar to the East for 2000 years, but not introduced to the West until the 13th century. The kite, used by the Chinese, manned for aerial observation and to test winds for sailing and unmanned as a signaling device and as a toy, held many of the answers to lifting a heavier-than-air device into the air.

One of the men who believed the study of kites unlocked the secrets of winged flight was Sir George Cayley. Born in England 10 years before the Montgolfier balloon flight, Cayley spent his 84 years seeking to develop a heavier-than-air vehicle supported by kite-shaped wings (Figure 1-2). The "Father of Aerial Navigation", Cayley discovered the basic principles on which the modern science of aeronautics is founded, built what is recognized as the first successful flying model, and tested the first full-size manned airplane.

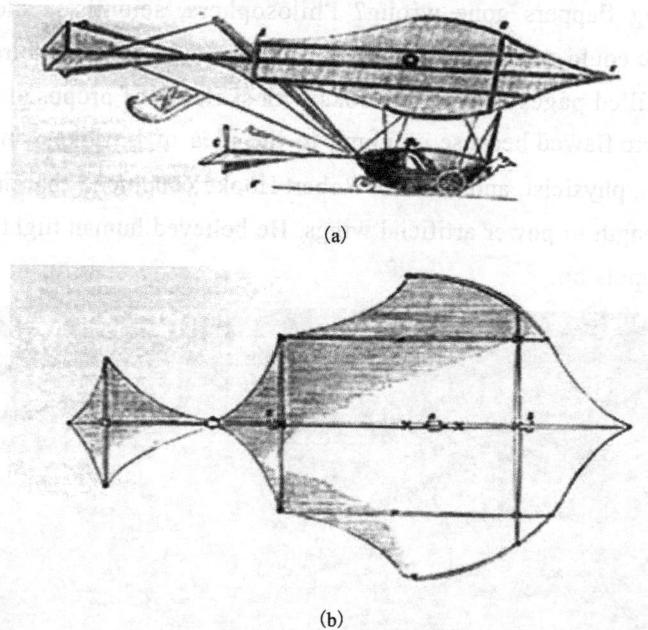

(a)

(b)

Figure 1-2 Glider from 1852 by Sir George Cayley, British aviator (1773–1857)

For the half-century after Cayley's death, countless scientists, flying enthusiasts, and inventors worked toward building a powered flying machine. Men, such as William Samuel Henson, who designed a huge monoplane that was propelled by a steam engine housed inside the fuselage and Otto Lilienthal, who proved mannned aircraft heavier than air was practical, worked towards the dream of powered flight. This dream was turned into reality by Wilbur and Orville Wright at Kitty Hawk, North Carolina, on December 17, 1903.

The bicycle-building Wright brothers of Dayton, Ohio, had experimented for 4 years with kites, their own homemade wind tunnel and different engines to power their biplane. One of their great achievements was proving the value of the scientific, rather than build-it-and-see approach to flight. Their biplane, The Flyer, combined inspired design and engineering with superior craftsmanship (Figure 1-3). By the afternoon of December 17,1903, the Wright brothers had flown a total of 98 seconds on four flights. The age of flight had arrived.

Figure 1-3 First flight by the Wright brothers

1.2 Introduction of Theories in the Production of Lift

The fundamental physical laws governing the forces acting upon an aircraft in flight were adopted from postulated theories developed before any human successfully flew an aircraft. The use of these physical laws grew out of the Scientific Revolution, which began in Europe in the 1600s. Driven by the belief that the universe operated in a predictable manner open to human understanding, many philosophers, mathematicians, natural scientists and inventors spent their lives attempting to unlock the secrets of the universe. One of the best known was Sir Isaac Newton, who not only formulated the Law of Universal Gravitation, but also described the three basic laws of motion.

In 1852, the German physicist and chemist, Heinrich Gustav Magnus (1802–1870), made experimental studies of the aerodynamic forces on spinning spheres and cylinders. (This had already been studied by Newton in 1672, in regard to spheres or tennis balls.) These experiments led to the discovery of the Magnus Effect, which helps explain the theory of lift.

A half-century after Newton formulated his laws, Daniel Bernoulli, a Swiss mathematician, explained how the pressure of a moving fluid (liquid or gas) varies with its speed of motion. Bernoulli's Principle states that as the velocity of a moving fluid (liquid or gas) increases, the pressure within the fluid decreases. This principle explains what happens to air passing over the curved top of the airplane wing.

Since air is recognized as a body and it is accepted that it must follow the above laws, one can begin to see how and why an airplane wing develops lift. As the wing moves through the air, the flow of air across the curved top surface increases in velocity creating a low-pressure area. Although Newton, Magnus, Bernoulli, and hundreds of other early scientists who studied the physical laws of the universe did not have the sophisticated laboratories available today, they provided great insight to the contemporary viewpoint of how lift is created.

The above content is only a brief overview of the theories in the production of lift, please see Chapter 3 for details.

1.3 Basic Engineering Design Process

The complete design process, from start to finish, is often outlined as in Figure 1-4. This iterative process begins with an identification of a need and a decision to something about it. After many iterations, the process ends with the presentation of the plan which will satisfy the need. Depending on the nature of the design task, several design phases may be repeated throughout the life of the product, from inception to termination. In the next several subsections, we shall examine these phases in detail.

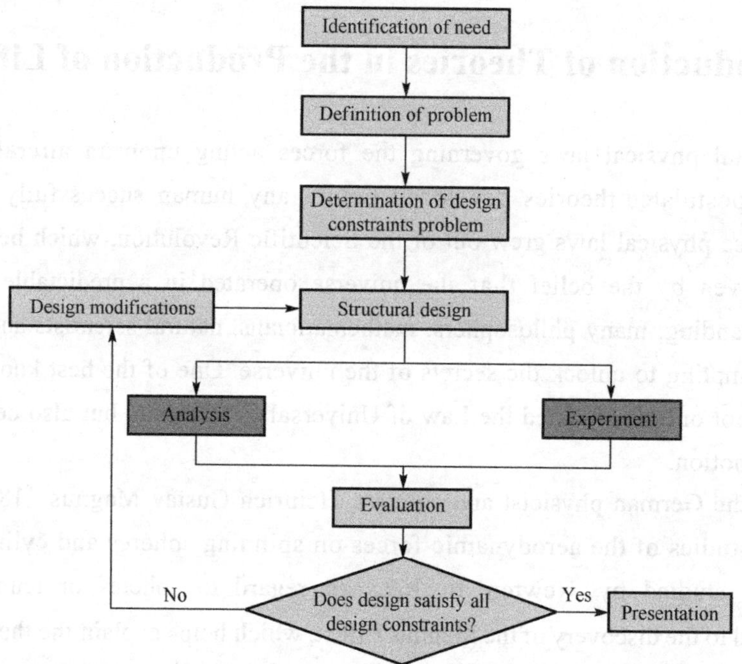

Figure 1-4 The design process

1.3.1 Identification of Need

Identification of the need is generally the first phase of the design process. It isn't always obvious that there is a need for a new design. This need may only arise from mild discontent or a sense that something is not right. The need is often not obvious at all. Recognition is often triggered by some particularly adverse occurrence or a set of random circumstances that arises almost simultaneously. For example, the need to do something about a food-packaging machine may be indicated by its noise level, by a variation in package weight, or by slight but perceptible variations in the quality of the packaging.

1.3.2 Definition of Problem

There is a distinct difference between the recogniton of a need and that of a problem. The definition of problem is more specific and must include all the specifications for the object that is to be designed. The specifications are the input and output quantities, the characteristics and dimensions of the space the object must occupy, and all the limitations on these quantities. We can regard the object to be designed as something in a black box. In this case we must specify the inputs and outputs of the box, together with their characteristics and limitations. The specifications determine the cost, the number to be manufactured, the expected life, the range, the operating temperature and the reliability. Specific characteristics can include the speeds, feeds, temperature limitations, maximum range, expected variations in the variables, dimensional and weight limitations, etc.

1.3.3 Design Constraint

There are many implied specifications that result either from the designer's particular environment or from the nature of the problem itself. The manufacturing processes that are available, together with the facilities of a certain plant, constitute restrictions on a designer's freedom, and hence are a part of the implied specifications. It may be that a small plant, for instance, does not own cold-working machinery. Knowing this, the designer might select other metal-processing methods that can be performed in the plant. The labor skills available and the competitive situation also constitute implied constraints. Anything that limits the designer's freedom of choice is a constraint. Many materials and sizes are listed in supplier's catalogs, for instance, but these are not all easily available and shortages frequently occur. Furthermore, inventory economics requires that a manufacturer stock a minimum number of materials and sizes.

1.3.4 Synthesis, Analysis and Optimization

The synthesis of a scheme connecting possible system elements is sometimes called the invention of the concept or concept design. This is the first and most important step in the synthesis task. Various schemes must be proposed, investigated and quantified in terms of established metrics. As the fleshing out of the scheme progresses, analyses must be performed to assess whether the system performance is satisfactory or better, and, if satisfactory, just how well it will perform. System schemes that do not survive analysis are revised, improved or discarded. Those with potential are optimized to achieve the best possible performance. Competing schemes are compared so that the path leading to the most competitive product can be chosen.

1.3.5 Mathematical Modeling and Evaluation

Both analysis and optimization require that we construct or devise abstract models of the system that will admit some form of mathematical analysis. We call these models mathematical models. In creating them it is our hope that we can find one that will simulate the real physical system very well. As indicated in Figure 1-4, evaluation is a significant phase of the total design process. Evaluation is the final proof of a successful design and usually involves the testing of a prototype in the laboratory. Here we wish to discover if the design really satisfies the needs. Is it reliable? Will it compete successfully with similar products? Is it economical to manufacture and to use? Is it easily maintained and adjusted? Can a profit be made from its sale or use? How likely is it to result in product-liability lawsuits? And is insurance easily and cheaply obtained? Is it likely that recalls will be needed to replace defective parts or systems?

1.3.6 The Presentation

Communicating the design to others is the final, vital presentation step in the design process. Undoubtedly, many great designs, inventions, and creative works have been lost to posterity simply because the originators were unable or unwilling to explain their accomplishments to others. Presentation is a selling job. The engineer, when presenting a new solution to administrative, management or supervisory persons, is attempting to sell or to prove to them that this solution is a better one. Unless this can be done successfully, the time and effort spent on obtaining the solution have been largely wasted. When designers sell a new idea, they also sell themselves. If they are repeatedly successful in selling ideas, designs, and new solutions to management, they begin to receive salary increases and promotions. In fact, this is how anyone succeeds in his or her profession.

1.4 Aircraft Types

1.4.1 Ultralight Vehicles

An ultralight aircraft (Figure 1-5) is referred to as a vehicle because the FAA(Federal Aviation Administration) does not govern it, if it:

(1) Is used or intended to be used by a single occupant.

(2) Is used for recreation or sport purposes.

(3) Does not have an airworthiness certificate.

(4) Is unpowered, weighing less than 155 pounds.

(5) Is powered, weighing less than 254 pounds empty weight, excluding floats and safety devices that are intended for deployment in a potentially catastrophic situation.

(6) Has a fuel capacity not exceeding 5 gallons.

(7) Is not capable of more than 55 knots calibrated airspeed at full power in level flight.

(8) Has a power-off stall speed, which does not exceed 24 knots calibrated airspeed.

Figure 1-5 A typical ultralight vehicle, which weighs less than 254 pounds

1.4.2 Light Sport Aircraft

In 2004, the FAA approved a new pilot certificate and aircraft category program to allow individuals to join the aviation community by reducing training requirements that affect the overall cost of learning to fly. The Sport Pilot Certificate was created for pilots flying light-weight, simple aircraft and offers limited privileges. The category of aircraft called the Light Sport Aircraft (LSA) includes airplane (land/sea), gyroplane, airship, balloon, weight-shift control (land/sea), glider, and powered parachute.

In order for an aircraft to fall in the Light Sport Category, it must meet the following criteria:

(1) The maximum gross takeoff weight may not exceed 1320 pounds, or 1430 pounds for seaplanes. Lighter-than-air maximum gross weight may not be more than 660 pounds.

(2) The maximum stall speed may not exceed 45 knots and the inflight maximum speed in level flight with maximum continuous power is not greater than 120 knots.

(3) Seating is restricted to single or two-seat configuration only.

(4) The power plant may be only a single, reciprocating engine (if powered), but may include rotary or diesel engines.

(5) The landing gear must be fixed, except gliders or those aircraft intended for operation on water.

(6) The aircraft can be manufactured and sold ready-to fly under a new special LSA category and certification must meet industry consensus standards. The aircraft may be used for sport, recreation, flight training and aircraft rental.

(7) The aircraft will have an FAA registration N-number and may be operated at night if the aircraft is properly equipped and the pilot holds at least a private pilot certificate with a minimum of a third-class medical.

Words and Expressions

1. The vehicle which can fly is classified as follows:

aircraft 飞行器

(1) lighter-than-air aircraft 轻于空气的飞行器

balloon 气球

airship 飞船

(2) rotorcraft 旋翼式飞行器

helicopter 直升机

gyroplane 旋翼机

(3) glider 滑翔机

(4) airplane 飞机

2. The airplane can be divided into the following groups according to the type of aircraft engine.

(1) propeller-driven aircraft 螺旋桨式飞机

piston propeller-driven aircraft 活塞螺旋桨式飞机

turboprop aircraft 涡轮螺旋桨式飞机

(2) jet aircraft 喷气式飞机

turbojet 涡轮喷气式飞机

turbofan jet 涡轮风扇喷气式飞机

3. The airplane can be divided into the following groups according to the aircraft's flight velocity.

(1) supersonic aircraft 超音速飞机

Supersonic aircraft travels faster than the speed of sound.

(2) hypersonic aircraft 高超声速飞机

Hypersonic aircraft travels much faster than the speed of sound.

(3) transonic aircraft 跨音速飞机

Transonic aircraft travels approximately equal to the speed of sound.

(4) subsonic aircraft 亚音速飞机

Subsonic aircraft travels slower than the speed of sound but also very fast.

4. The airplane can be divided into the following groups according to the aircraft's air range.

(1) short-range aircraft / short-haul aircraft 近程飞机

The air range is shorter than 2400 kilometers.

(2) medium-haul aircraft 中程飞机

The air range is longer than 2400 kilometers but shorter than 4800 kilometers.

(3) long-range aircraft / long-haul aircraft 远程飞机

The air range is longer than 4800 kilometers.

5. The airplane can be divided into the following groups according to the types of landing gears.

(1) landplane 陆上飞机

A landplane is an airplane that can take off from or land on land.

(2) seaplane / hydroplane 水上飞机

A seaplane or hydroplane is a type of airplane that can take off from or land on water.

(3) amphibian 水陆两用飞机

An amphibian is an airplane which can land on both land and water.

6. The airplane can be divided into the following groups according to the number of wings.

(1) monoplane 单翼飞机

A monoplane is an airplane with a single pair of wings.

(2) biplane 双翼飞机

A biplane is an old-fashioned type of airplane with two pairs of wings, one above the other.

(3) multiplane 多翼飞机

A multiplane is an airplane with more than two pairs of wings, say, three or four pairs.

7. The airplane can be divided into the following groups according to the positions of the wing relative to the fuselage.

(1) low wing airplane 下单翼飞机

The airplane whose wing is below the cabin.

(2) mid wing airplane 中单翼飞机

The airplane whose wing is anywhere between the fuselage.

(3) high wing airplane 上单翼飞机

The airplane whose wing is above the cabin.

8. The airplane can be divided into the following groups according to its mission.

(1) war plane 军用飞机

(2) civil aircraft 民用飞机

第1章 简 介

1.1 航 空 历 史

随着时间流逝，人们早已忘记谁是第一批给自己装上翅膀，从悬崖跳下想飞起来的“鸟人”，但是他们失败的经历可以为那些想飞起来的人提供参考。“翅膀”到底哪里出了问题？哲学家、科学家和发明家都提供了解决方案，但没有人能够给人安装翅膀，使人像鸟类一样飞翔。16 世纪，列奥纳多·达·芬奇在他的笔记本上记录了很多他构思的飞行器草图，但是他的大部分想法存在缺陷，因为他坚持模仿鸟类翅膀(图 1-1)。到 1655 年，数学家、物理学家兼发明家罗伯特·胡克得出结论：人类的身体没有足够力量去驱动人造翅膀。他相信人类飞行需要某种形式的人造推进器。

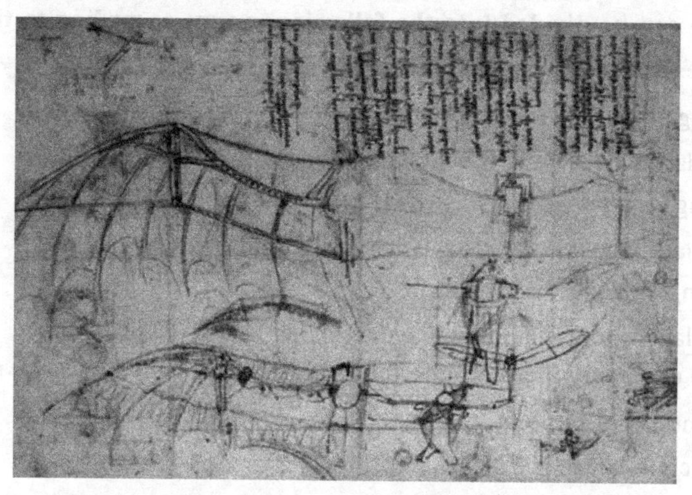

图 1-1 达·芬奇设计的扑翼机

人类对飞行的渴望促使人们在其他方面进行实践。1783 年，由约瑟夫和艾蒂安·蒙戈菲尔制作的第一个载人热气球在空中飞行了 23 分钟。十天后，雅克·查尔斯教授驾驶了历史上第一个气球。气球飞行的疯狂行为，激发了人们的想象力，在很长一段时间，飞行爱好者们将自己的专业转向研究实现轻于空气的飞行上面。

气球解决了飞行中的升力问题，但这只是人类飞行中的问题之一。如何控制速度和方向却难住了气球飞行研究者。人们最终在一个东方国家熟悉了 2000 年的儿童玩具中找到了解决这个问题的办法，但是这个东西直到 13 世纪才传入西方。风筝被中国人用来进行载人的空中观察、航行中的测风，也用作不载人的信号装置，还可作为一个玩具，它的飞行原理可以用来解释为什么比空气重的东西可以升起来。

乔治·凯利爵士认为对风筝的研究可以揭开有翼飞行的秘密。在蒙戈菲尔气球飞行的十年前，乔治·凯利在英格兰出生，他用一生的时间发明了具有风筝形状翅膀且比空气重的飞行器(图 1-2)。作为“航空之父”，凯利发现了现代航空科学建立的基本原理，创建了公认的第一个成功的飞行模型，并测试了第一架全尺寸载人飞行器。

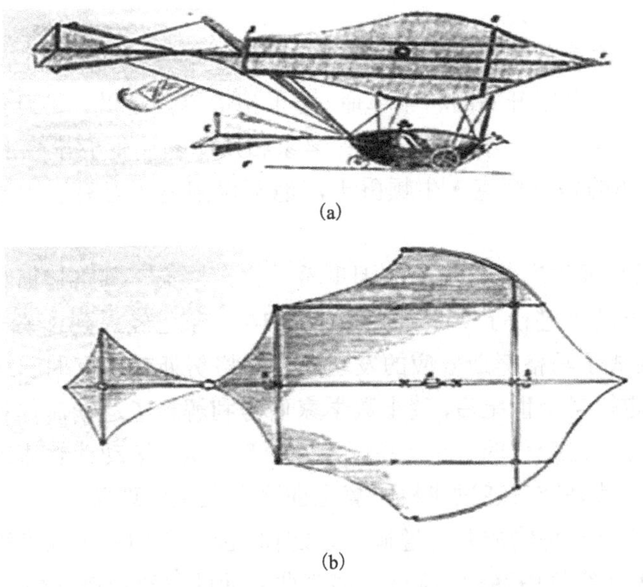

(a)

(b)

图 1-2　1852 年滑翔机，发明者为乔治·凯利爵士，英国飞行员(1773—1857)

在凯利死后的半个世纪里，无数的科学家、飞行爱好者和发明家致力于建造动力飞行器。比如威廉·塞缪尔·汉森，他设计了一个依靠机身里的蒸汽机推动的巨大单翼机，奥托利·利连撒尔证明了人类驾驶比空气重的飞行器进行飞行是可行的，他们两人都为实现动力飞行的梦想而努力。1903 年 12 月 17 日，北卡罗来纳州的基蒂霍克，威尔伯和奥维尔·莱特将这一梦想变成了现实。

发明自行车的莱特兄弟(代顿，俄亥俄州)，在 4 年的时间里用风筝和自制风洞做了大量的实验，他们用不同的引擎来驱动双翼飞机。莱特兄弟的伟大成就之一是证明科学的价值，而不是建造飞行器和看待飞行的方式。他们启发性地将飞行器设计和高超的工艺相结合，设计出名为“飞行者”的双翼飞行器(图 1-3)。截止到 1903 年 12 月 17 日下午，莱特兄弟设计的飞机在 4 次起降中共飞行了 98 秒。飞行时代到来了。

图 1-3　莱特兄弟的第一次飞行

1.2　升力产生理论概述

在人类飞机试飞成功之前，可用于解释飞机飞行的物理学理论就已经建立。这些物理定律的使用源于欧洲十七世纪开始的科学革命。由于相信宇宙是以一种可预见和人们可以理解的方式运行，许多哲学家、数学家、自然科学家和发明家用他们毕生的精力试图揭开宇宙的奥妙。最广为人知的是艾萨克·牛顿爵士，他不仅阐述了万有引力定律，也描述了运动的三个基本定律。

1852 年，德国物理学家兼化学家海因里希·古斯塔夫·马格努斯(1802—1870)，对旋转球体和圆柱体的气动力进行了实验研究。(牛顿 1672 年已经提到过关于球体或网球的这种效应。)这些实验促成了马格努斯效应的发现，而马格努斯效应有助于解释上升理论。

牛顿提出运动定律半个世纪后，瑞士数学家伯努利解释了运动流体(液体或者气体)的压力是如何随其运动速度而变化的。伯努利压力差原理认为，运动或者流动的速度增加会导致流体压力的降低。这就是空气通过飞机机翼上曲面所发生的现象。

由于空气被认为是一种物质并且遵循上述定律，因此我们就可以了解机翼产生升力的原因和原理。随着机翼在空气中移动，流过上部弯曲表面的气流速度加快，并形成一个低压区。尽管牛顿、马格努斯、伯努利和成百上千的其他早期的科学家们研究宇宙的物理规律时没有像今天一样有可用的复杂的实验室，但是他们对当代升力的产生理论提供了巨大的帮助。

以上内容仅对升力产生的原理做了简单的概述，具体的内容详见第 3 章。

1.3　基本工程设计过程

从开始到结束的完整设计过程，如图 1-4 所示。这个迭代过程以识别需求和决定具体操作开始。经过多次迭代，这个过程以满足需求计划的提出而结束。根据设计任务的性质，从开始到结束，在产品的整个生命周期几个设计阶段可能会重复。接下来，我们将在设计过程中详细讨论这些步骤。

1.3.1　识别需求

设计过程一般从识别需求开始。新设计的需求并不总是显而易见的。需求在刚开始可能只是一种说不清的不满意或者似乎不太对的感觉。需求经常是不明显的。认识通常是由一个特定的恶劣环境或几乎同时出现一组随机的情况触发的。例如，在制作食品包装机时，我们需要根据噪音水平、包装重量的变化，以及在包装或包裹时的轻微但可观察到的质量变化来决定怎么做。

1.3.2　定义问题

需求的识别和对问题的定义之间有一个明显的区别。问题的定义更具体，必须包括所有设计对象的规格。规格有输入量和输出量，对象必须占据的空间的特性和尺寸，以及其他所有对这些量的限制。我们可以把对象设计成一个装在黑盒子里的东西。在这种情况下，我们

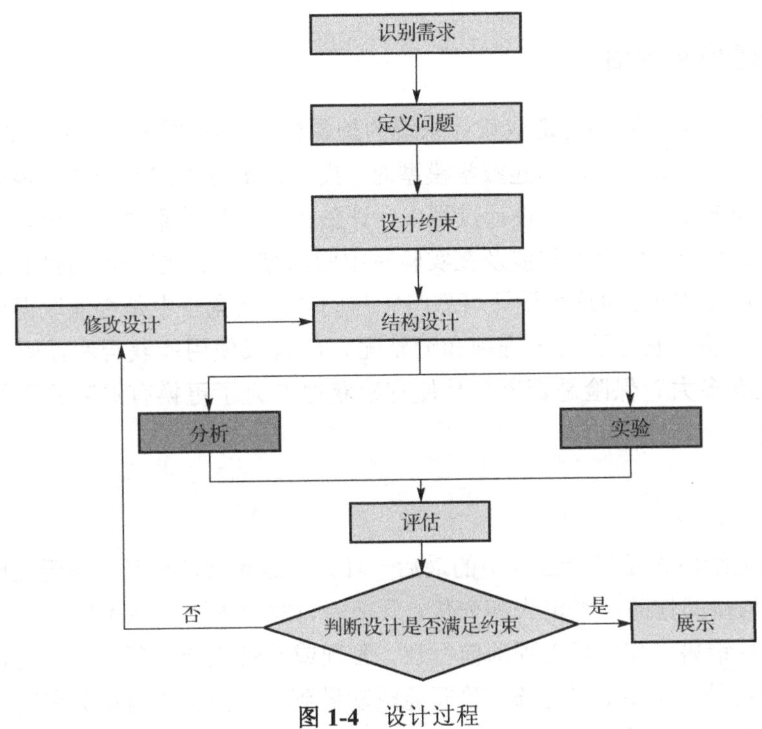

图1-4 设计过程

必须明确盒子的输入和输出，以及它们的特点和局限性。规范定义了成本、生产数量、预期寿命、范围、操作温度和可靠性。指定的特性可包括速度、食物、温度的限制、最大范围、预期中的变量的变化、尺寸和重量的限制等。

1.3.3 设计约束

许多隐含的规格源自设计师所在的特定环境，或者问题本身的性质。可以采用的制造过程与特定工厂的设备，共同构成对设计师自由的限制，并成为隐含规格的一部分。例如，这可能是一个小厂并且没有自己的冷加工设备。知道了这一点，设计者就可能选择其他的可在工厂进行的金属加工方法。现有的劳动技能和竞争形势也构成了隐含的限制。凡是制约设计师自由选择的限制，都是一种约束。例如，许多供应商目录中列出的材料和尺寸，这些不容易获取，并且经常发生短缺现象。此外，库存经济学要求制造商储存少量材料。

1.3.4 合成、分析与优化

计划的合成是对可能的系统元素的连接，有时也被称为概念发明或概念设计。这是合成任务的第一步，也是最重要的一步。参照既定指标，提出各种方案，进行研究并量化。在对方案进行具体化工作时，我们必须进行分析，以便评估系统的性能是否令人满意或更好，并且在满意的基础上分析其执行的效果。没有通过分析的系统方案要修订、改进或放弃，而有潜力的方案要进行优化设计，以便获得最佳的性能。对具有竞争力的方案进行比较，以便选择一个能生产出具有竞争力产品的方案。

1.3.5 数学建模和评估

分析和优化都需要我们建造或设计系统的抽象模型，以便进行某种形式的数学分析。我们称这些模型为数学模型。在创建数学模型时，我们希望可以找到一个能很好模拟真实物理系统的模型。如图 1-4 所示，在设计过程中，评估是一个极其重要的部分。评估将最终证明我们的设计是否成功，而且通常涉及在实验室中的原型测试。这一步的目的是我们希望发现设计是否真正满足需求。如这个设计可靠吗？与同类产品竞争中是否能取得成功？它在生产和使用中是否经济？它是否易于维护和调整呢？出售或使用能获得利益吗？导致产品责任诉讼的可能性有多大？保险是否容易且便宜地获得？为了更换有缺陷的零件或系统可能需要召回产品吗？

1.3.6 展示

将设计传达给他人是设计过程中的最后一环，也是重要的一环。毫无疑问，许多伟大的设计、发明和创作的作品已经消失和失传，只是因为发起人不能或不愿给别人解释他们的成就。展示是一种销售工作。当工程师向行政、管理或监管人展示新的方案时，他正试图出售或向他们证明这是一个更好的方案。除非能成功做到这一点，否则花费在获得解决方案上的时间和精力将会被大大浪费。当设计师推销一个新的想法时，他们也在推销自己。当设计者不断成功地向管理人员推销自己的新想法、新设计和新的解决方法时，他们将收获加薪和升职。事实上，这也是人们在他或她的职业生涯中取得成功的秘诀。

1.4 飞 机 种 类

1.4.1 超轻型飞行器

超轻飞行器(图 1-5)作为一种交通工具，如果满足以下条件则不受 FAA(美国联邦航空管理局)监管：

(1)只能被一个所有者使用或者打算使用。

图 1-5 典型的重量低于 254 磅的飞行器

(2) 用于娱乐或者体育运动。

(3) 没有适航证书。

(4) 非机动的，重量低于 155 磅(1 磅=0.453592 千克)。

(5) 如果是机动的，空载时重量须低于 254 磅，不包括用于潜在灾难情况下的救生圈和安全设备的重量。

(6) 有一个容量不超过 5 加仑 (1 加仑(美)=3.78543 升) 的燃料箱。

(7) 全功率水平飞行时校正空速不能超过 55 节(1 节=1.852 千米/小时)。

(8) 动力系统关闭时的失速速度不超过 24 节。

1.4.2　轻型运动飞行器

2004 年，美国联邦航空管理局批准了一项新的飞行员证书和飞机类别计划，允许个人通过减少影响总体学习成本的培训要求加入航空界。运动飞行员证书是为那些驾驶质量轻的简易飞行器的驾驶员设立的，但是这种证书只提供有限的特权。轻型运动飞行器的种类包括飞机(水陆两用)、旋翼机、飞艇、热气球、重量转移控制飞行器(水陆两用)、滑翔机、机动跳伞。

轻型运动飞行器必须满足以下条件：

(1) 最大起飞重量不超过 1320 磅，对于水上飞机不超过 1430 磅。比空气轻的航天器最大总重量不超过 600 磅。

(2) 最大失速速度不超过 45 节，持续动力下水平飞行时的最大速度不超过 120 节。

(3) 只有一个或两个座位。

(4) 动力装置只有一个单独的往复式发动机，包括转缸式发动机或者柴油发动机。

(5) 起落架是固定的，除了滑翔机或者用于水上作业的飞行器之外。

(6) 属于新型特殊轻型运动飞行器种类的飞行器能被制造、销售和飞行，并且认证必须符合行业共识的标准。这类飞行器可以用于运动、娱乐、飞行训练和租赁。

(7) 这类飞行器将会在 FAA 注册一个 N 号，如果这类飞行器装备合适，并且飞行员至少持有一个最低三级的私人飞行员医疗证明，就可以在夜间飞行。

Chapter 2 Aircraft Structure

An aircraft is a machine that is able to fly by gaining support from the air, or, in general, the atmosphere of a planet. It counters the force of gravity by using either static lift or by using the dynamic lift of an airfoil, or in a few cases the downward thrust from the engine. Although airplanes are designed for a variety of purposes, most of them have the same major components (Figure 2-1). The overall characteristics are largely determined by the original design objectives. Most airplane structures include a fuselage, wings, an empennage, landing gear, and a powerplant.

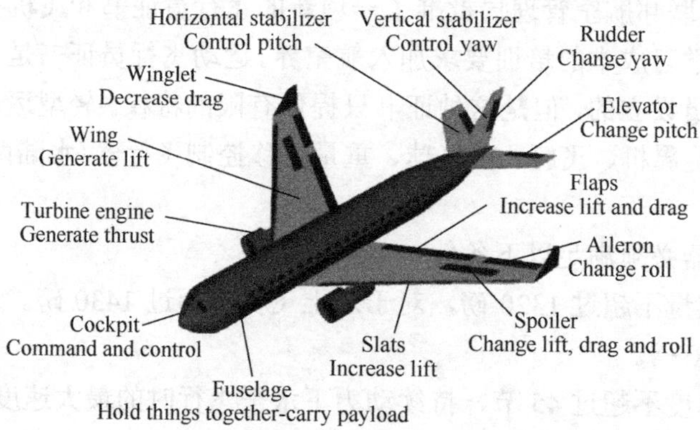

Figure 2-1 Major components

2.1 Types of Aircraft Construction

2.1.1 Truss Structure

The main drawback of truss structure is its lack of a streamlined shape. In this construction method, lengths of tubing, called longerons, are welded in place to form a well-braced framework. Vertical and horizontal struts are welded to the longerons that give the structure a square or rectangular shape when viewed from the end. Additional struts are needed to resist stress that can come from any direction. Stringers and bulkheads or formers, are added to shape the fuselage and support the covering. As technology progressed, aircraft designers began to enclose the truss members to streamline the airplane and improve performance. This was originally accomplished with cloth fabric, which eventually gave way to lightweight metals such as aluminum. In some cases, the outside skin can support all or a major portion of the flight loads. Most modern aircrafts use a form of this stressed skin structure known as monocoque or semimonocoque construction (Figure 2-2).

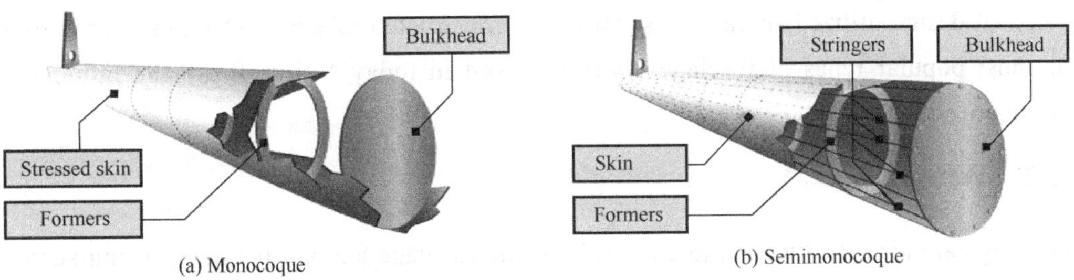

(a) Monocoque (b) Semimonocoque

Figure 2-2 monocoque and Semimonocoque fuselage design

2.1.2 Monocoque

Monocoque construction uses stressed skin to support almost all loads much like an aluminum beverage can. Although very strong, monocoque construction is not highly tolerant to deformation of the surface. For example, an aluminum beverage can support considerable forces at the ends of the can, but if the side of the can is deformed slightly while supporting a load, it collapses easily. Because most twisting and bending stresses are carried by the external skin rather than by an open framework, the need for internal bracing was eliminated or reduced, saving weight and maximizing space. Although employed in the early aviation period, monocoque construction would not reemerge for several decades due to the complexities involved. Every day examples of monocoque construction can be found in automobile manufacturing where the unibody is considered standard in manufacturing.

2.1.3 Semimonocoque

Semimonocoque construction, partial or one-half, uses a substructure to which the airplane's skin is attached. The substructure, which consists of bulkheads and/or formers of various sizes and stringers, reinforces the stressed skin by taking some of the bending stress from the fuselage. The main section of the fuselage also includes wing attachment points and a firewall. On single-engine airplanes, the engine is usually attached to the front of the fuselage. There is a fireproof partition between the rear of the engine and the flight deck or cabin to protect the pilot and passengers from accidental engine fires. This partition is called a firewall and is usually made of heat-resistant material such as stainless steel. However, a new emerging process of construction is the integration of composites or aircraft made entirely of composites.

2.2 Major Components

2.2.1 Fuselage

The fuselage is the central body of an airplane and is designed to accommodate the crew, passengers, and cargo. It also provides the structural connection for the wings and tail assembly. Older types of aircraft design utilized an open truss structure constructed of wood, steel, or aluminum tubing. The most popular types of fuselage structures used in today's aircraft are the monocoque and semimonocoque.

2.2.2 Wings

The wings are airfoils attached to each side of the fuselage and are the main lifting surfaces that support the airplane in flight. There are numerous wing designs, sizes, and shapes used by the various manufacturers. Each fulfills a certain need with respect to the expected performance for the particular airplane. Wings may be attached at the top, middle, or lower portion of the fuselage. These designs are referred to as high-, mid- and low-wing, respectively. The number of wings can also vary. Airplanes with a single set of wings are referred to as monoplanes, while those with two sets are called biplanes.

The principal structural parts of the wing are spars, ribs, and stringers (Figure 2-3). These are reinforced by trusses, I-beams, tubing or other devices, including the skin. The wing ribs determine the shape and thickness of the wing.

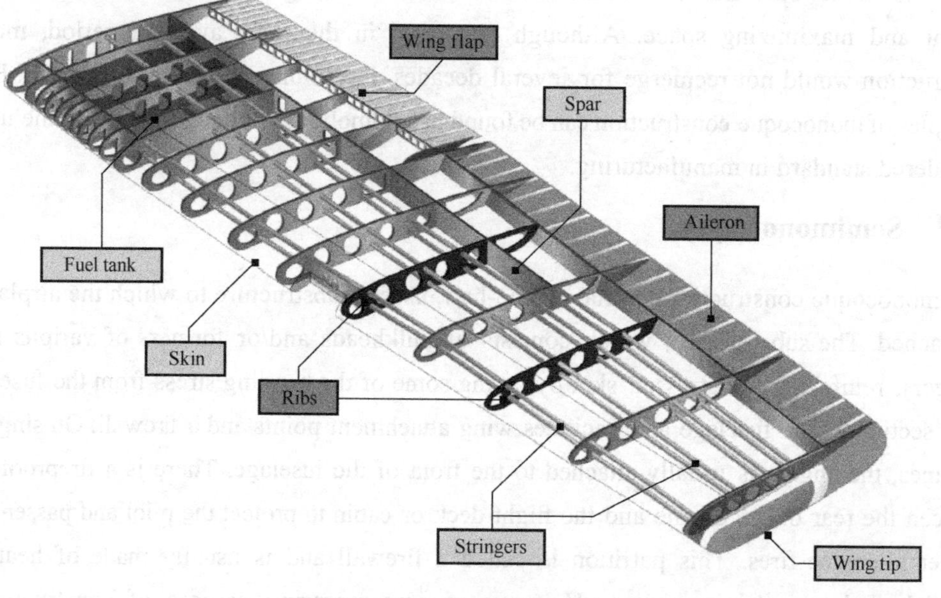

Figure 2-3 Wing components

2.2.3 Empennage

The empennage includes the entire tail section and consists of fixed surfaces such as the vertical stabilizer and the horizontal stabilizer. The movable surfaces include the rudder, the elevator and one or more trim tabs (Figure 2-4).

The rudder is attached to the back of the vertical stabilizer. During flight, it is used to move the airplane's nose left and right. The elevator, which is attached to the back of the horizontal stabilizer, is used to move the nose of the airplane up and down during flight. Trim tabs are small, movable portions of the trailing edge of the control surface. These movable trim tabs, which are controlled from the flight deck, reduce control pressures. Trim tabs may be installed on the ailerons, the rudder and/or the elevator.

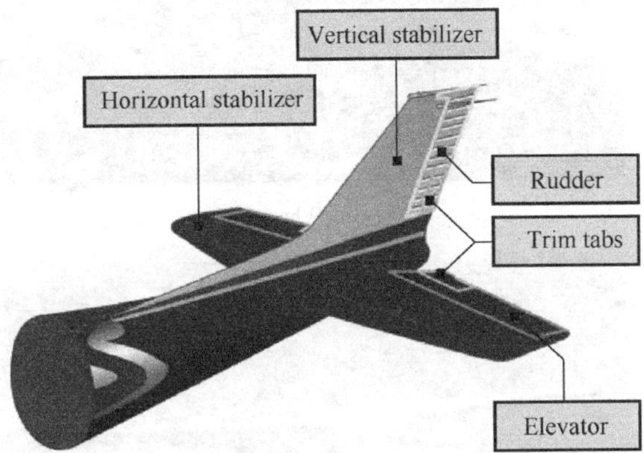

Figure 2-4 Empennage Components

A second type of empennage design does not require an elevator. Instead, it incorporates a one-piece horizontal stabilizer that pivots from a central hinge point. This type of design is called a stabilator and is moved using the control wheel, just as the elevator is moved. For example, when a pilot pulls back on the control wheel, the stabilator pivots so the trailing edge moves up. This increases the aerodynamic tail load and causes the nose of the airplane to move up. Stabilators have an antiservo tab extending across their trailing edge (Figure 2-5).

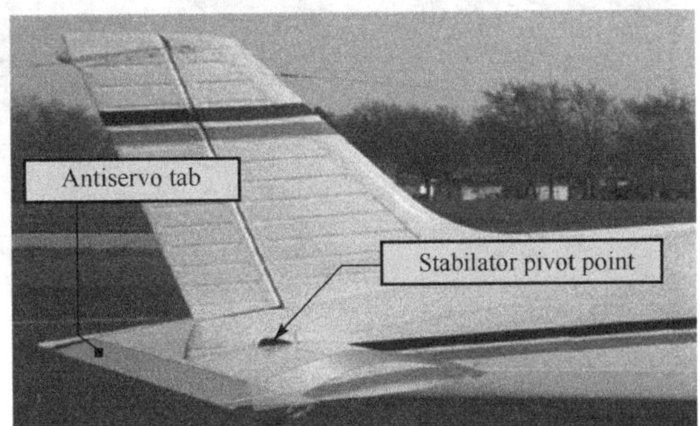

Figure 2-5 Stabilator components

2.2.4 Landing Gear

The landing gear is the principal support of the airplane when parked, taxiing, taking off or landing. The most common type of landing gear consists of wheels, but airplanes can also be equipped with floats for water operations or skis for landing on snow (Figure 2-6).

The landing gear consists of three wheels—two main wheels and a third wheel positioned either at the front or rear of the airplane. Landing gear with a rear mounted wheel is called conventional landing gear.

(a) Wheels

(b) Floats

(c) Skis

Figure 2-6 Types of landing gear

Airplanes with conventional landing gear are sometimes referred to as tailwheel airplanes. When the third wheel is located on the nose, it is called a nosewheel and the design is referred to as a tricycle gear. A steerable nosewheel or tailwheel permits the airplane to be controlled throughout all operations while on the ground. Most aircraft are steered by moving the rudder pedals, whether nosewheel or tailwheel. Additionally, some aircraft are steered by differential braking.

2.2.5 The Powerplant

The powerplant usually includes both the engine and the propeller. The primary function of the engine is to provide the power to turn the propeller. It also generates electrical power, provides a vacuum source for some flight instruments and in most single-engine airplanes, provides a source of heat for the pilot and passengers (Figure 2-7). The engine is covered by a cowling or a nacelle, which are both types of covered housings. The purpose of the cowling or nacelle is to streamline the flow of air around the engine and to help cool the engine by ducting air around the cylinders.

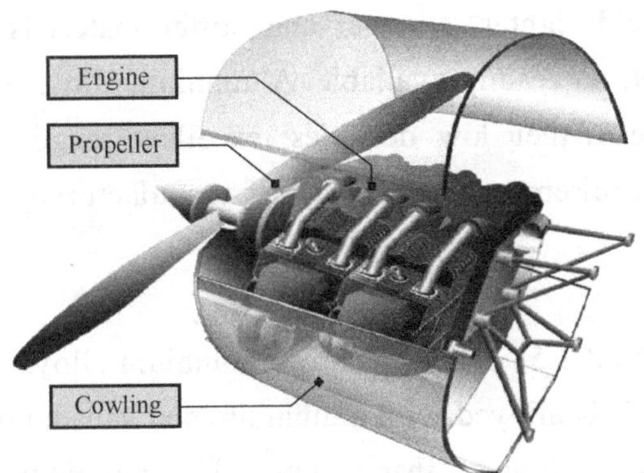

Figure 2-7 Engine compartment

The propeller, mounted on the front of the engine, translates the rotating force of the engine into thrust, a forward acting force that helps move the airplane through the air. The propeller may also be mounted on the rear of the engine as in a pusher-type aircraft. A propeller is a rotating airfoil that produces thrust through aerodynamic action. A low pressure area is formed at the back of the propeller's airfoil and high pressure is produced at the face of the propeller, similar to the way lift is generated by an airfoil used as a lifting surface or wing. This pressure differential pulls air through the propeller, which in turn pulls the airplane forward.

2.3 Aircraft Structure Materials

To some extent the structural layout of an airframe is dependent upon the choice of the material of construction. It is necessary to be aware of the available alternatives before the structural design is undertaken. In some circumstances it is desirable to consider different designs employing alternative materials and to compare them before making a final decision on the material to be used. The primary drivers in airframe design are low weight and low cost. These two criteria have to be balanced in order to achieve an acceptable solution. Both are a function of the material itself and the manufacturing processes used to make the details and assemblies of a structural component. Thus the best solution would be an airframe constructed of an expensive, but low density material if the associated manufacturing costs are lower than those associated with a cheaper material or if the impact of low weight on the life cycle costs of the aircraft is dominant.

2.3.1 Classification of Materials

1. Metallic Materials

1) Aluminum Alloys

In order to obtain a suitable airframe material, aluminum is alloyed with various other metals such as magnesium, although zinc and copper are more common. Developments using lithium as an alloying metal yield lighter, stronger and stiffer materials but they are much more expensive and not, as yet, so readily available. Aluminium alloys are the most commonly used airframe material because of their low densities, excellent range of properties which can be matched to particular requirement, experience of manufacturing techniques and the known in-service behavior.

2) Titanium Alloys

Titanium has a density about 1.58 times that of aluminium alloys. Some applications use it in pure form but most often it is alloyed with aluminium and vanadium. The strength/density ratio is, if anything, somewhat higher than comparable aluminium alloys but the elastic modulus/density ratio is rather less. The strength properties are retained at much higher temperatures than light alloy and the metal is less susceptible to corrosion.

3) Steel

Steel is available with a wide range of properties and can have greater strength than either light alloys or titanium alloys. Its density is some 2.8 times that of aluminium and as a consequence the main use of steel is for highly loaded components where it is desirable to keep the size as compact as possible. Some types of steel are corrosion resistant.

2. Fiber-reinforced Composite Materials

A composite consists of two distinct elements which, unlike metal alloys, remain separate within the material. In airframe applications one of the elements is a fibrous material used to reinforce the other, a matrix material. Although the fibers can be short and random they are more usually long and arranged directionally to impart the required strength properties in specific directions. The most frequently used reinforcing fibers are:

(1) Carbon (graphite);

(2) Glass;

(3) Aramid, such as Kevlar.

2.3.2 Criteria for the Selection of Materials

For airframe application the most important material properties are:

(1) elastic modulus (E) and shear modulus (G);

(2) ductile yield or proof strength (σ_b) and ultimate tensile strength (UTS);

(3) fracture toughness (K_{IC}), which influences the brittle strength;

(4) density (ρ).

For a given material these properties will vary with environmental conditions, such as operating temperature, and their relative importance depends upon the form of the loading and application. For an airframe the most important loading forms are:

(1) tension, including the effect of pressure in thin shells;

(2) buckling of slender columns;

(3) buckling of plate-like components (including reinforced plates);

(4) bending of plate-like components.

2.3.3 Application of Materials

1. Metals

When sheets of metal are subjected to pressure and so-called "super-plastic" temperatures they may be given very large deformations. This technique is especially valuable in the application of thin material since the shaping process may be used to confer local stiffness.

Metallic components are commonly joined using mechanical fasteners such as rivets and bolts, but welding and bonding offer alternatives in appropriate circumstances. Bonding may use special glues or a diffusion technique. Diffusion bonding of two components occurs when the temperature is sufficient for diffusion mechanisms to operate across the contact areas. In some cases, for example titanium, the super-plastic and diffusion temperatures are similar such that separate pans may be joined at the same time as the super-plastic forming. When this is the case the process is very efficient and versatile, although the production numbers have to be sufficient to justify the high tooling costs. While aluminium alloys may also be super-plastically formed, diffusion bonding is less readily achieved due to the existence of oxide coatings on the surfaces of the metal. They do not dissolve at high temperature as is the case with titanium.

1) Aluminium Alloys

Aluminium alloys are by far the most widely used airframe materials and have been developed to meet specific airframe applications. Indeed, until the relatively recent introduction of composites, it was true to say that, with very few exceptions, all airframes were mainly of light alloy. Aluminium alloys have very good strength to density ratios, especially in the compression and bending of relatively thin plates and sections. A thin outer cladding of pure aluminium can confer good corrosion resistance but this does have an adverse effect upon fracture toughness and hence it is not now used as much as it was at one time. Low strength aluminium-magnesium alloys can be welded by conventional techniques and higher grade copper-based alloys may be spot-welded. Recent developments in friction-stir welding (FSW) are also finding application in comparatively substantial light alloy assemblies.

2) Titanium Alloys

Although the application of titanium alloys is not extensive, partly because of high cost, they do possess advantages making their choice appropriate in some circumstances:

(1) The retention of properties at comparatively high temperatures, which is useful for engine fire bulkheads and other applications where high temperature is a major consideration.

(2) The coefficient of thermal expansion is low and similar to that of carbon-reinforced plastic material. It is appropriate to use titanium fittings where concentrated loads need to be reacted in carbon-fiber-reinforced structures.

(3) Certain titanium alloys can be super-plastically formed and simultaneously diffusion bonded. This process is suitable for complex components.

Otherwise titanium alloys are mainly used where the high loading intensity and confined space render the use of aluminium alloys inefficient. Titanium bolts are also often used.

3) Steel

Usually only high-strength versions of steel are used and like titanium it is particularly applicable when loading intensity is very high. Experience has shown that it is difficult to design a complete airframe in steel efficiently. When loading and high temperature rule out aluminium alloys and composites, titanium is the preferred material. Steel is normally limited in application to detail fittings and components of joints, including bolts, together with large specific items such as landing gear parts.

2. Composites

Apart from potentially high material properties an advantage of fiber-reinforced composites is the ability to tailor the properties to meet a given set of requirements. Effectively a specific material can be developed for each application and complex shapes can be readily moulded. However, against these major attributes the more intensive quality control considerations and the general problem of joining parts without losing the potential improvements must be set. There is also a need to ensure that the assembly has satisfactory electromagnetic properties, which may mean including conducting fibers or providing an external conducting mesh. The facility for the repair of damage is another issue demanding consideration.

In practice composite materials and components may be formed in a number of ways. In addition to the embedding of fibers in a matrix it is possible to produce "ply"-type materials consisting of laminates of thin metal and reinforced plastic. Similarly sandwich construction may be employed with metal or fiber-reinforced plastic faceplates and a variety of core materials. Both of these arrangements have advantages in producing stiff materials, often with good acoustic damping and related properties.

A form of composite, applicable to machine parts, is the reinforcement of the matrix with short fibers or particles. Silicon carbide fibers find application in materials intended for high-temperature powerplant applications using metals, such as titanium, or a ceramic as the matrix. The titanium composites are produced by hot isostatic pressing or diffusion bonding of fibers and foil while vapour deposition is used with ceramic matrices.

Words and Expressions

1. Airframe 飞机机架

(1) fuselage 机身

The fuselage (or body) of the airplane holds all of the parts together and carries the passengers or cargo.

(2) wing / airfoil 机翼

The wings generate most of the lift necessary for flight.

① aileron 副翼

Ailerons are parts of the wing which are used to roll the aircraft. At the trailing edge, it's an aileron.

② leading edge slat 前缘缝翼

③ trailing edge flap 后缘襟翼

Flaps and slats are parts of the wing which are used during taking off and landing to increase lift and drag.

④ spoiler 扰流板

Spoilers are parts of the wing which are used to roll the aircraft and to decrease lift and increase drag during landings. Between the leading and trailing edges, it's a spoiler.

(3) empennage / the tail assembly 尾翼

① vertical stabilizer 垂直尾翼

② rudder 方向舵

A rudder is a control surface connected to the vertical stabilizer and used to yaw the aircraft.

③ horizontal stabilizer 水平尾翼

④ elevator 升降舵

Elevators are control surfaces connected to the horizontal stabilizer and are used to pitch the aircraft.

⑤ stabilator 全动式水平尾翼

Stabilators are control surfaces which provide both horizontal stability and pitch control for the aircraft.

(4) landing gear / undercarriage 起落架

Airplanes require landing gear for taxiing, takeoff and landing.

(5) powerplant 动力装置

Most modern aircrafts are powered by gas turbines, or jet engines.

2. Fuselage Construction 机身构造(Figure 2-8)

cabin / passenger cabin 客舱

cargo / cargo compartment　货舱

cockpit/flight compartment　驾驶舱

bulkhead　隔板

former　隔框

wing attachment points　机翼安装处

firewall　防火墙

stringer　长桁

stressed skin　耐压蒙皮

semi-monocoque　半硬壳式结构

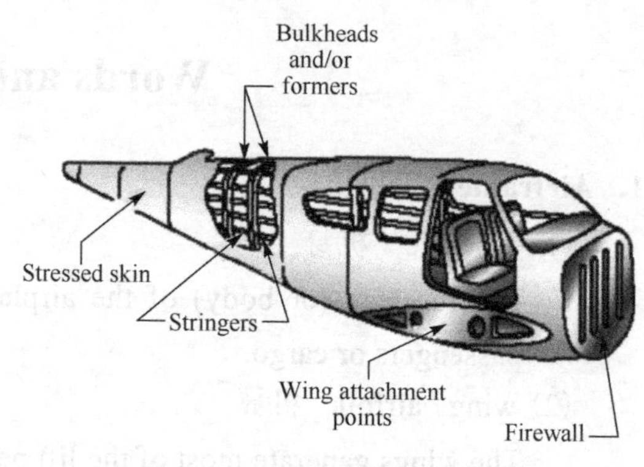

Figure 2-8

3. **Wing Components　机翼组件**(Figure 2-9)

wing tip　翼尖

rib　翼肋

spar　翼梁

fuel tank　油箱

wing flap　襟翼

stringer　桁条

skin　蒙皮

aileron　副翼

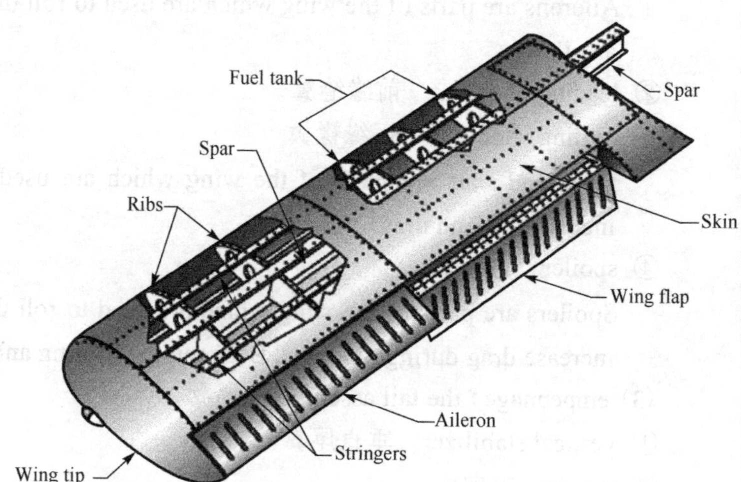

Figure 2-9

4. **Empennage Components　尾翼组件**(Figure 2-10)

vertical stabilizer　垂直尾翼

rudder　方向舵

horizontal stabilizer　水平尾翼

elevator　升降舵

stabilator　全动式水平尾翼

trim tab　配平调整片

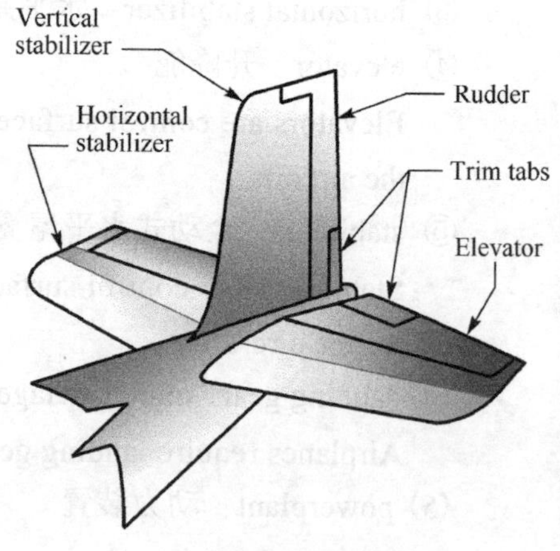

Figure 2-10

5. **Landing Gear Components　起落架组件**(Figure 2-11)

nose landing gear　前起落架

main landing gear　主起落架

wheel (起落架)轮

conventional landing gear　传统起落架

tail wheel　机尾(起落架)轮

tricycle landing gear　三轮式起落架

nose wheel　机头(起落架)轮

6. Powerplant Compartment 发动机室(Figure 2-12)

engine 发动机，引擎

propeller 螺旋桨

cowling/nacelle 引擎罩，整流罩

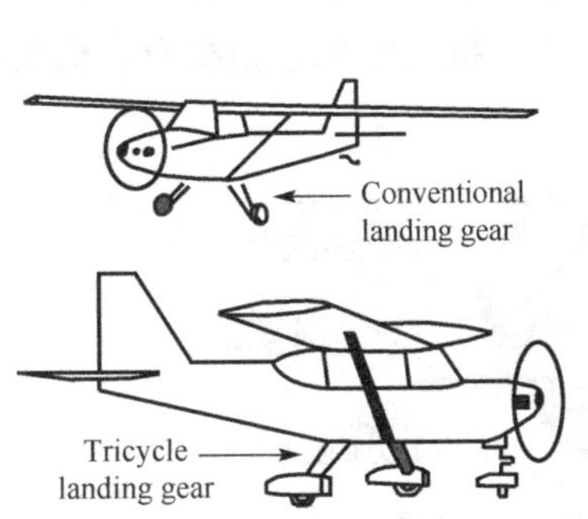

Figure 2-11

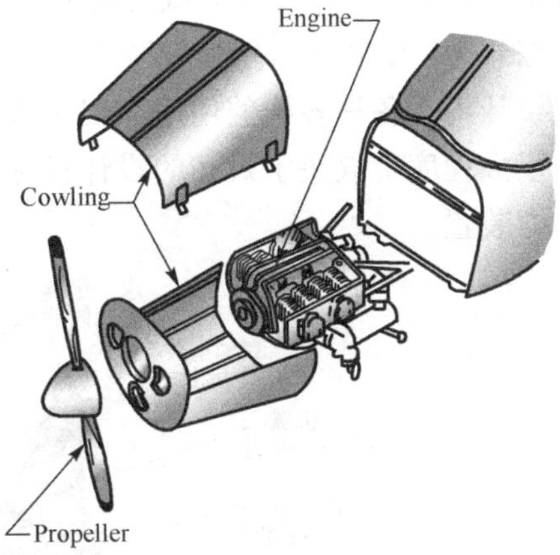

Figure 2-12

7. Fastener 紧固件

structural joint 结构连接点

bolt 螺栓，螺钉

rivet 铆钉

swage 接头锁紧螺母

nut 螺母，螺帽

screw 螺丝

pin 销钉

cotter 开口销

washer 垫圈

hinge 铰链

wrench 扳钳

bearing 轴承

shank 轴

8. Implement 工具

milling machine 铣床

boring machine 镗床

punching machine 冲床

turning machine 车床

sawing machine 锯床

routing machine 特形铣床

第 2 章　飞 机 结 构

飞机是一种能靠空气支持，或者说靠大气层支持来飞行的机器。不管机翼产生的静态升力或者动态升力，或者在一些情况下由发动机产生的向下推力，飞机用它们来克服重力。虽然飞机被设计用于各种各样的用途，但是大多数飞机具有相同的主要组件(图 2-1)。它的总体特性在很大程度上取决于最初的设计目标。大多数飞机的结构包括机身、机翼、尾翼、起落架和动力装置。

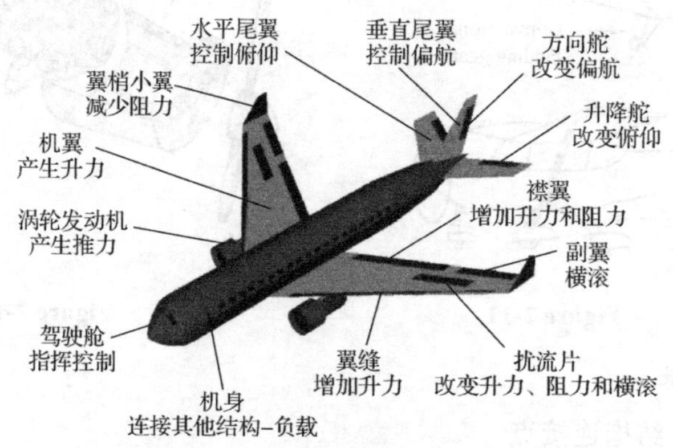

图 2-1　飞机机架的主要组成部分

2.1　飞机结构类型

2.1.1　桁架结构

桁架结构的主要缺点是缺乏流线型外形。在这种构造方法中，纵梁靠焊接连接，形成一个稳固的支撑框架。把垂直和水平框架焊接到纵梁，从尾部观察将看到纵梁成正方形或矩形。抵制来自任意方向的压力要用其他框架。纵梁、隔框或框架共同构成机身并用来支撑覆盖物。随着技术进步，飞行器设计人员开始把桁架单元围起来，使之成为流线型的飞机以改进性能。最初使用布料织物来实现这一目标，最终换用轻质金属，如铝。在某些情况下，蒙皮可以支持所有或者主要部分的飞行载荷。大多数现代飞机使用这种应力蒙皮结构形式称为硬壳式或半硬壳式结构，如图 2-2 所示。

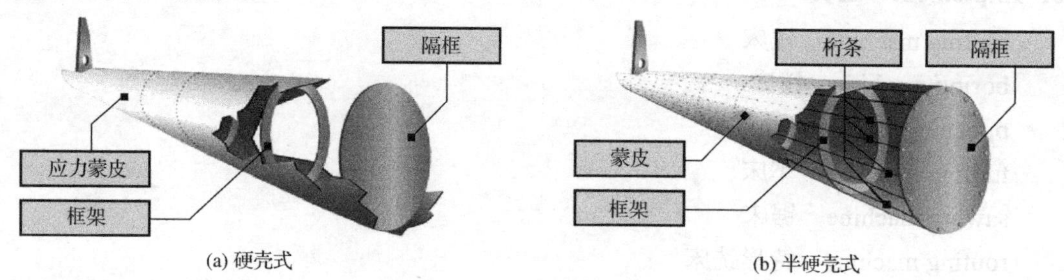

图 2-2　硬壳式或半硬壳式机身设计

2.1.2　硬壳式

硬壳构造设计使用应力蒙皮来支持几乎全部的载荷，它很像铝制饮料罐。尽管这种结构非常结实，但是表面不能有凹痕或者变形。这种特性可以很容易通过一个铝制饮料罐来演示。当对饮料罐的两边施加相同的力时，饮料罐不会损坏。然而，如果罐壁上有一点凹痕，那么这饮料罐子就很容易扭曲变形。因为大部分的扭曲和弯曲应力是靠蒙皮而不是框架来承载的，从而对内部支撑的需求会消失或减少，这样就可以减少重量增大内部空间。虽然早期的航空设计中就采用了硬壳式结构，但是由于其复杂性，往后几十年再也没出现过。汽车制造业中有使用硬壳式结构，这种结构在汽车生产过程中被视为一种标准化。

2.1.3　半硬壳式

半硬壳式结构部分或半部分使用飞机蒙皮可以贴附的亚结构，亚结构由舱隔和各种不同尺寸的防水隔层以及桁条组成，通过来自机身的弯曲应力来加固应力蒙皮。机身的主要部分也包括机翼挂载点和防火隔板。在单引擎飞机上，引擎一般装在机身的前端。在引擎后面、驾驶舱和飞行甲板或客舱之间用防火区来保护飞行员和乘客，以免受到引擎火焰的伤害。这个区称为防火墙，一般由阻热材料如不锈钢制成。但是新结构是复合材料的集成或者完全是由复合材料构成。

2.2　主要组成部分

2.2.1　机身

机身是飞机的中央部位，用来承载机组人员、乘客和货物。在结构上，它连接机翼和机尾。过去的飞机设计利用木、钢或铝管组成开放式桁架结构。现在最受欢迎的机身结构是硬壳式结构或者半硬壳式结构。

2.2.2　机翼

机翼是由翼型组成连接到机身两侧的结构，是支持飞机飞行的主要升力面。很多飞机制造商设计了多种不同的机翼样式、尺寸和外形。每一种都是为了满足特定的需求，这些需求由具体飞机的目标性能决定。机翼可以安装在机身的上、中或较低部分，分别称为上单翼、中单翼和下单翼设计。机翼的数量也可以不同，有一组机翼的飞机称为单翼飞机，有两组机翼的飞机称为双翼飞机。

机翼的主要结构部件有翼梁、翼肋和桁条(图 2-3)。这些都通过支杆、I 字形梁、管子或其他结构包括蒙皮而加固。翼肋决定了机翼的外形和厚度。

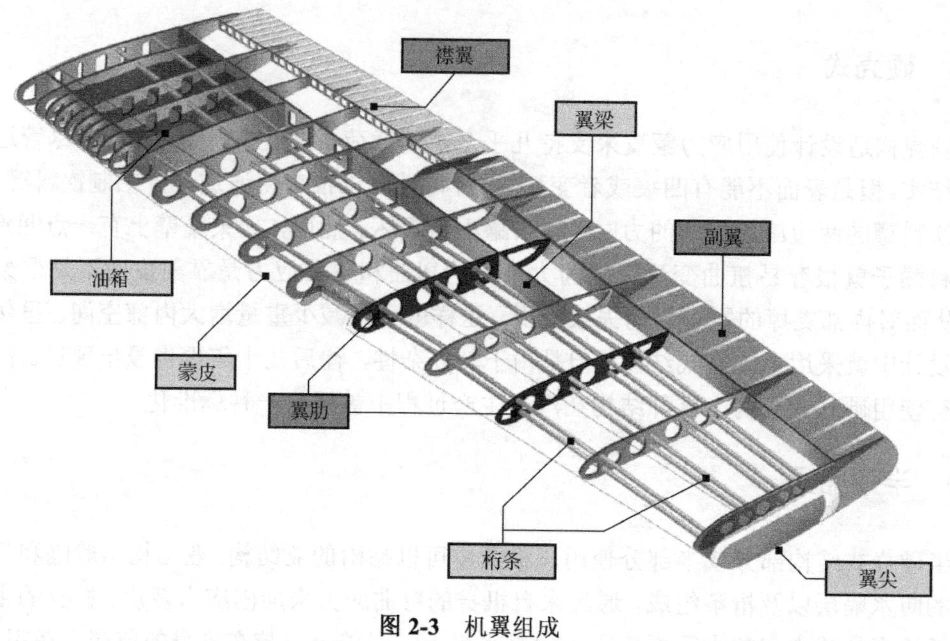

图 2-3 机翼组成

2.2.3 尾翼

尾翼包括整个尾部部分，由固定翼面如垂直尾翼和水平尾翼组成。可活动的表面包括方向舵、升降舵和一个或者多个配平片(图 2-4)。

方向舵安装在垂直尾翼的后部。飞行时，它能使飞机头部向左或者向右运动。升降舵安装在水平尾翼的后面，用于控制在飞行中飞机的头部向上或者向下运动。配平片是位于控制面的尾部边缘较小可活动的部分。这些可活动的由驾驶舱控制的配平片，可降低控制压力。配平片可以安装在副翼、方向舵或升降舵上。

另一种尾翼的设计不需要升降舵。相反，在中央的铰链点安装一片水平尾翼，铰链轴是水平的。这种类型的设计叫全动式水平尾翼，使用控制轮移动达到升降舵的效果。例如，当飞行员向后拉控制轮时，水平尾翼转动，于是拖尾边缘向上运动。这增加了空气动力学尾部载荷并导致飞机机头向上移动。水平尾翼还有一个沿尾部边缘的防沉降片，如图 2-5 所示。

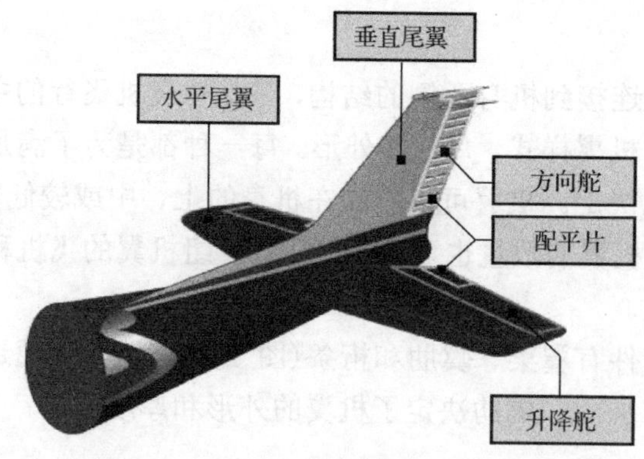

图 2-4 尾翼组成

图 2-5　全动式水平尾翼组成

2.2.4　起落架

起落架是飞机停放、滑行、起飞或者着陆时的主要支撑部分。大多数起落架由轮子组成，但是飞机也可以安装浮筒以便在水上操作，或者安装用于雪上着陆的雪橇，起落架的类型如图 2-6 所示。

(a) 轮式

(b) 漂浮式

(c) 雪橇式

图 2-6　起落架的类型

起落架由三个轮子组成，即两个主轮和一个安装在飞机后面或者前面的副轮。在后部安装第三个轮子的起落架称为传统起落架。

具有传统起落架的飞机称为后三点式飞机。第三个轮子安装在飞机头部位置的称为前三点式飞机，相应的这种设计叫三轮车式起落架。可操控的前轮或者尾轮允许在地面上通过各种操作来控制飞机。大部分飞机的方向操控是靠移动方向舵脚踏板、前轮或者尾轮。除此之外，一些飞机靠差动制动。

2.2.5 动力装置

动力装置一般包括发动机和螺旋桨。发动机的主要作用是为螺旋桨提供转动的动力。发动机也产生电力，为一些飞行仪器提供真空源，在大多数单发动机飞机上，为飞行员和乘客提供热量的来源，如图 2-7 所示。发动机由飞机引擎罩盖住，或者放置在飞机引擎机舱里。引擎罩和引擎机舱是两种盖套。引擎罩或者引擎机舱的作用是使发动机周围的空气流动变得顺利，通过引导气缸周围的空气来冷却发动机。

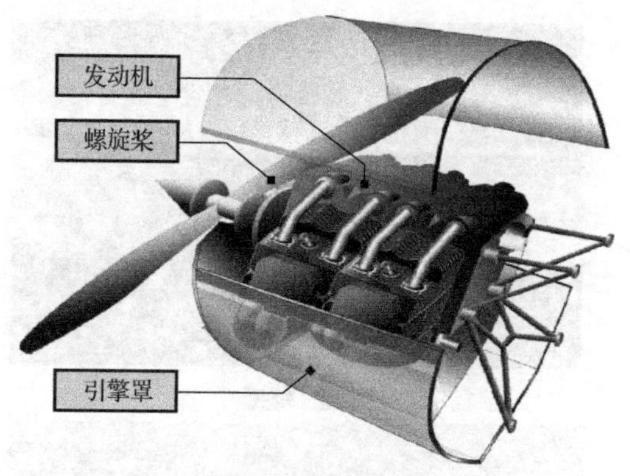

图 2-7 发动机室

安装在发动机前部的螺旋桨将发动机的旋转力转化为推力，这是一种向前的作用力使飞机在空中前行。螺旋桨也可以安装在发动机的后部推动飞机前进。螺旋桨是通过空气动力作用产生推力的旋转式机翼。在螺旋桨翼面的背部形成低压，前面形成高压，类似于机翼被用作升力面或升力翼时产生的升力。这个压差迫使空气通过螺旋桨反过来推动飞机前进。

2.3 飞机结构材料

机体的结构布局某种程度上取决于其所选用的结构材料。因此，在结构设计之前有必要了解各种可选择的材料。在某些情况下，需要考虑不同设计所选用的材料，比较各种材料的特性以便最终选择一种最好的材料。在机身设计中最主要的考虑指标是重量小和成本低。这两个指标需要相互妥协平衡以达到最终可接受的方案。材料本身的功能和制造工艺常常决定了结构件的细节和装配性。因此，如果制造成本比使用廉价材料带来的制造成本低，或者低重量对飞机寿命周期中（使用、维护等）成本起到主导影响，那么最好的解决办法就是用一种昂贵但密度小的材料建造机身。

2.3.1 材料的分类

1. 金属材料

1) 铝合金

为了获得一种合适的机身材料，一般采用铝镁合金或其他铝合金，尽管锌和铜更常见。使用锂合金可以获得更轻、更硬的材料，但是它们的价格昂贵，也不易获得。铝合金是最常用的机身材料，因为其密度低，性能可以满足特殊要求，制造工艺成熟，并具有众所周知的服役特性。

2) 钛合金

钛合金的密度是铝合金的 1.58 倍。某些应用中会使用纯钛金属，但常见的是使用它与铝和钒构成的合金。钛合金的强度密度比高于铝合金，但是弹性模量密度比低于铝合金。与轻合金相比，钛合金的强度特性在高温条件下仍然得以保持，而且具有很强的耐腐蚀性。

3) 钢

钢特性范围很广，比其他轻合金或钛合金强度大。钢的密度是铝的 2.8 倍，它最主要的应用是制造高载荷组件，从而尽可能保持紧凑的尺寸。某些类型的钢也具有耐腐蚀性。

2. 纤维增强复合材料

复合材料由两种不同的元素组成，不同于金属合金，复合材料中的组成元素在材料中保持分离。在机身中使用纤维材料来加固基础材料，尽管纤维材料本身很短和随机，但是它们常常在指定的方向上长距离排列，以传递载荷。最常用的增强纤维材料有：

(1) 碳(石墨)；

(2) 玻璃；

(3) 芳族聚酰胺纤维，如凯芙拉。

2.3.2 材料选择的标准

对于机身而言，最重要的材料特性是：

(1) 弹性模量(E)和剪切模量(G)；

(2) 材料韧性或者抗拉强度(σ_b)和极限拉伸强度(UTS)；

(3) 影响脆性强度的断裂韧性(K_{IC})；

(4) 密度(ρ)。

对于一种特定的材料，它的材料特性会随着环境条件的变化而变化，如使用温度，材料特性的相对重要性依赖于应用场合和载荷形式。对于机身而言，最重要的载荷形式是：

(1) 张力，包括薄壳受到的压力影响；

(2) 梁的屈曲；

(3) 板结构的屈曲(包括加强筋)；

(4) 板结构的弯曲。

2.3.3 材料应用

1. 金属

当金属板进行压力试验和所谓的“超塑性”温度试验时会发生很大的形变。这项技术在薄材料使用中非常有价值，因为它的成形方法可以用于加强局部刚度。

金属部件通常采用机械紧固件如铆钉和螺栓来连接，但在合适的场合中可以采用焊接和粘接替代。粘接可使用特殊的胶黏剂或扩散技术。当温度足够时，接触面的扩散机制将会作用，从而进行两个组件的扩散接合。在一些情况下，以钛为例，它的“超塑性”温度和扩散温度相近，因而在“超塑”成形的同时发生分离。这种情况下制造工艺是非常高效的，尽管要有足够的产品数量来覆盖高成本加工。而铝合金也可以“超塑”成形，由于金属表面有氧化物，使得扩散接合不易实现。这些材料与钛一样在高温下不能溶解。

1) 铝合金

铝合金是目前使用最广泛的机身材料，它可以用来制造在特定情况下应用的机身。事实上，直到近些年才引用复合材料，但是除了特殊情况外，几乎所有的机身都由轻合金制造而成。铝合金具有很好的比强度，特别是在薄壁结构的压缩和弯曲。纯铝具有很强的耐腐蚀性，但它在断裂韧性上有不利因素，因此现在不像以前那么经常使用纯铝了。可以使用低强度的铝镁合金进行传统技术焊接，用高品质的铜基合金进行点焊。近年来，搅拌摩擦焊接技术(FSW)的发展使其用于轻合金的焊接。

2) 钛合金

由于钛合金成本较高，所以钛合金应用并不广泛，但是它在某些情况下具有很大的优势：

(1) 钛合金在高温下可以保持稳定的材料特性，在航空发动机和其他需要考虑高温的环境中非常有用。

(2) 类似于碳纤维材料，钛合金的热膨胀系数较低。钛合金可以在载荷集中区域使用。

(3) 钛合金可以被“超塑”成形，同时扩散接合。这一工艺非常适合制造复杂组件。

钛合金主要用于高载荷密度区域。钛合金螺栓也经常被使用。

3) 钢

通常只使用高强度钢，如钛合金一般只用在高载荷密度区域。经验表明，很难设计一个高效的全钢材料的完整机身结构。当高载荷和高温度下不适宜使用铝合金和复合材料时，钛则是首选材料。钢通常只在有限的场合下使用，例如用于组件连接的接头(如螺栓)，用于连接大部件(如起落架)。

2. 复合材料

除了潜在的高性能外，纤维增强复合材料的优势在于能够定制材料特性以满足一系列既定的要求。可以根据每一种应用，有效地开发其特定的材料，并且可以容易地塑造出复杂的形状。但是定制材料的特性必须有更加严格的质量控制，在不失去材料潜在提升性能的前提下要考虑连接件之间普遍存在的问题。此外，还需要保证这些组件具有满意的电磁特性，这意味着包含一些导电纤维或者提供外部导电网络。设备的可修复性也是一个需要考虑的问题。

在实践中，复合材料可以通过多种方式形成。除此之外，可以在一定厚度的薄金属和增强塑料中嵌入纤维形成“ply”型的材料。同样，三明治夹层结构可能采用金属或者纤维增强塑料和各种芯材。这两种方法都有助于制造出很硬的材料，通常具有很好的声学阻尼特性和其他相关特性。

有一种复合材料可用于制造机械零件，它是一种短纤维或颗粒增强材料。碳化硅纤维被应用于制造高温动力装置，它用金属(如钛或者陶瓷)作为基体。钛复合材料可以通过热压或者扩散接合的方法将纤维材料和金属箔复合，或者与陶瓷基体进行蒸汽沉积。

Chapter 3　Knowledge of Flight

This chapter examines the knowledge of flight, including the fundamental physical laws governing the forces acting on an aircraft in flight, and what effect these natural laws and forces have on the performance characteristics of aircraft, meanwhile the principles of flight and flight controls. To control an aircraft, be it an airplane, helicopter, glider, or balloon, the pilot must understand the principles involved and learn to use or counteract these natural forces.

3.1　Structure of the Atmosphere

The atmosphere is an envelope of air that surrounds the Earth and rests upon its surface. It is as much a part of the Earth as its land and water. However, air differs from land and water inasmuch as it is a mixture of gases. It has mass, weight, and indefinite shape.

Air, like any other fluid, is able to flow and change its shape when subjected to even minute pressures because of the lack of strong molecular cohesion. For example, gas will completely fill any container into which it is placed, expanding or contracting to adjust its shape to the limits of the container. The atmosphere is composed of 78 percent nitrogen, 21 percent oxygen, and 1 percent other gases, such as argon or helium. Most of the oxygen is contained below 35000 ft altitude.

3.1.1　Atmospheric Pressure

Though there are various kinds of pressure, pilots are mainly concerned with atmospheric pressure. It is one of the basic factors in weather changes, helps to lift the aircraft, and actuates some of the most important flight instruments in the aircraft. These instruments often include the altimeter, the airspeed indicator（ASI）, the vertical speed indicator, and the manifold pressure gauge.

Though air is very light, it has mass and is affected by the attraction of gravity. Therefore, like any other substance, it has weight; because it has weight, it has force. Since it is a fluid substance, this force is exerted equally in all directions, and its effect on bodies within the air is called pressure. Under standard conditions at sea level, the average pressure exerted by the weight of the atmosphere is approximately 14.7 pounds per square inch（psi）. The density of air has significant effects on the aircraft's performance. As air becomes less dense, it reduces:

(1) Power, because the engine takes in less air.

(2) Thrust, because the propeller is less efficient in thin air.

(3) Lift, because the thin air exerts less force on the airfoils.

The pressure of the atmosphere varies with time and altitude. Due to the changing atmospheric pressure, a standard reference was developed. The standard atmosphere at sea level is a surface temperature of 59 degrees Fahrenheit（℉）or 15 degrees Celsius（℃）and a surface pressure of 29.92 inches of mercury（in Hg）or 1013.2 millibars（mbar)（Figure 3-1）.

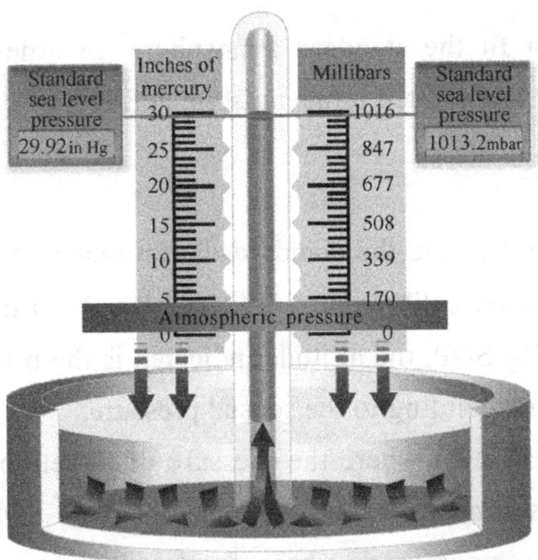

Figure 3-1 Standard sea level pressure

A standard temperature lapse rate is one in which the temperature decreases at the rate of approximately 3.5℉ or 2℃ per thousand feet up to 36000ft. Above this point, the temperature is considered constant up to 80000ft. A standard pressure lapse rate is one in which pressure decreases at a rate of approximately 1in Hg per 1000ft of altitude gain to 10000ft (Figure 3-2).

The International Civil Aviation Organization (ICAO) has established this as a worldwide standard, and it is often referred to as International Standard Atmosphere (ISA) or ICAO Standard Atmosphere. Any temperature or pressure that differs from the standard lapse rates is considered nonstandard temperature and pressure. Adjustments for nonstandard temperatures and pressures are provided on the manufacturer's performance charts.

Altitude/ft	Pressure /in Hg	Temperature	
		/℃	/℉
0	29.92	15.0	59.0
1000	28.86	13.0	55.4
2000	27.82	11.0	51.8
3000	26.82	9.1	48.4
4000	25.84	7.1	44.8
5000	24.89	5.1	41.2
6000	23.98	3.1	37.6
7000	23.09	1.1	34.0
8000	22.22	−0.9	30.4
9000	21.38	−2.8	27
10000	20.57	−4.8	23.4
11000	19.79	−6.8	19.8
12000	19.02	−8.8	16.2
13000	18.29	−10.8	12.6
14000	17.57	−12.7	9.1
15000	16.88	−14.7	5.5
16000	16.21	−16.7	1.9
17000	15.56	−18.7	−1.7
18000	14.94	−20.7	−5.3
19000	14.33	−22.6	−8.7
20000	13.74	−24.6	−12.3

Figure 3-2 Properties of standard atmosphere

Since all aircraft performance is compared and evaluated with respect to the standard atmosphere, all aircraft instruments are calibrated for the standard atmosphere. Thus, certain corrections must apply to the instrumentation, as well as the aircraft performance, if the actual operating conditions do not fit the standard atmosphere. In order to account properly for the nonstandard atmosphere, certain related terms must be defined.

3.1.2 Pressure Altitude

Pressure altitude is the height above the standard datum plane (SDP). The aircraft altimeter is essentially a sensitive barometer calibrated to indicate altitude in the standard atmosphere. If the altimeter is set for 29.92in Hg SDP, the altitude indicated is the pressure altitude—the altitude in the standard atmosphere corresponding to the sensed pressure.

The SDP is a theoretical level where the pressure of the atmosphere is 29.92in Hg and the weight of air is 14.7 psi. As atmospheric pressure changes, the SDP may be below, at, or above sea level. Pressure altitude is important as a basis for determining aircraft performance, as well as for assigning flight levels to aircraft operating at above 18000ft.

The pressure altitude can be determined by either of two methods:

(1) By setting the barometric scale of the altimeter to 29.92in Hg and reading the indicated altitude, or

(2) By applying a correction factor to the indicated altitude according to the reported "altimeter setting".

3.1.3 Density Altitude

The more appropriate term for correlating aerodynamic performance in the nonstandard atmosphere is density altitude—the altitude in the standard atmosphere corresponding to a particular value of air density.

Density altitude is pressure altitude corrected for nonstandard temperature. As the density of the air increases (lower density altitude), aircraft performance increases. Conversely, as air density decreases (higher density altitude), aircraft performance decreases. Density altitude is used in calculating aircraft performance. Under standard atmospheric condition, air at each level in the atmosphere has a specific density; pressure altitude and density altitude identify the same level. Density altitude, then, is the vertical distance above sea level in the standard atmosphere at which a given density is to be found.

The computation of density altitude must involve consideration of pressure (pressure altitude) and temperature. Since aircraft performance data at any level are based upon air density under standard day conditions, such performance data apply to air density levels that may not be identical to altimeter indications. Under conditions higher or lower than standard, these levels cannot be determined directly from the altimeter.

Density altitude is determined by first finding pressure altitude, and then correcting this altitude for nonstandard temperature variations. Since density varies directly with pressure, and inversely with temperature, a given pressure altitude may exist for a wide range of temperature by allowing the density to vary. However, a known density occurs for any one temperature and pressure altitude. The density of the air, of course, has a pronounced effect on aircraft and engine performance. Regardless of the actual altitude at which the aircraft is operating, it will perform as though it were operating at an altitude equal to the existing density altitude.

For example, when set at 29.92in Hg, the altimeter may indicate a pressure altitude of 5000ft. According to the AFM/POH, the ground run on takeoff may require a distance of 790ft under standard temperature conditions.

However, if the temperature is 20℃ above standard, the expansion of air raises the density level. Using temperature correction data from tables or graphs, or by deriving the density altitude with a computer, it may be found that the density level is above 7000ft, and the ground run may be closer to 1000ft.

Air density is affected by changes in altitude, temperature, and humidity. High density altitude refers to thin air while low density altitude refers to dense air. The conditions that result in a high density altitude are high elevations, low atmospheric pressures, high temperatures, high humidity, or some combination of these factors. Lower elevations, high atmospheric pressure, low temperatures, and low humidity are more indicative of low density altitude.

Using a flight computer, density altitude can be computed by inputting the pressure altitude and air temperature outside at flight level. Density altitude can also be determined by referring to the table and chart in Figure 3-3 and Figure 3-4.

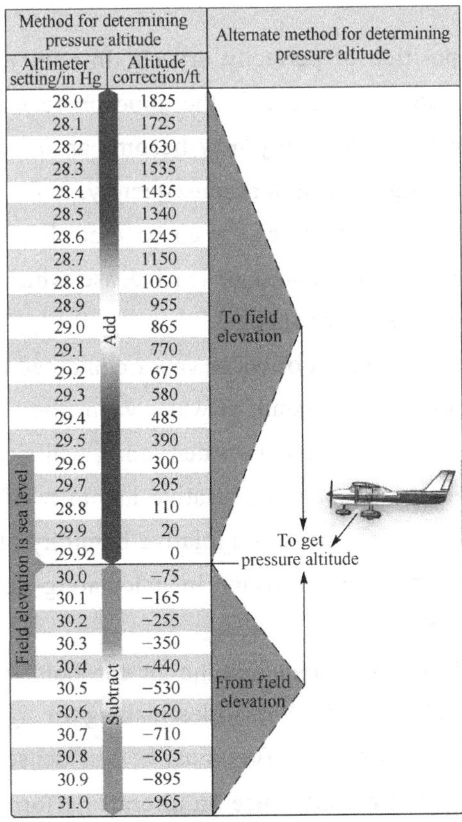

Altimeter setting/in Hg	Altitude correction/ft
28.0	1825
28.1	1725
28.2	1630
28.3	1535
28.4	1435
28.5	1340
28.6	1245
28.7	1150
28.8	1050
28.9	955
29.0	865
29.1	770
29.2	675
29.3	580
29.4	485
29.5	390
29.6	300
29.7	205
28.8	110
29.9	20
29.92	0
30.0	−75
30.1	−165
30.2	−255
30.3	−350
30.4	−440
30.5	−530
30.6	−620
30.7	−710
30.8	−805
30.9	−895
31.0	−965

Figure 3-3 Field elevation versus pressure

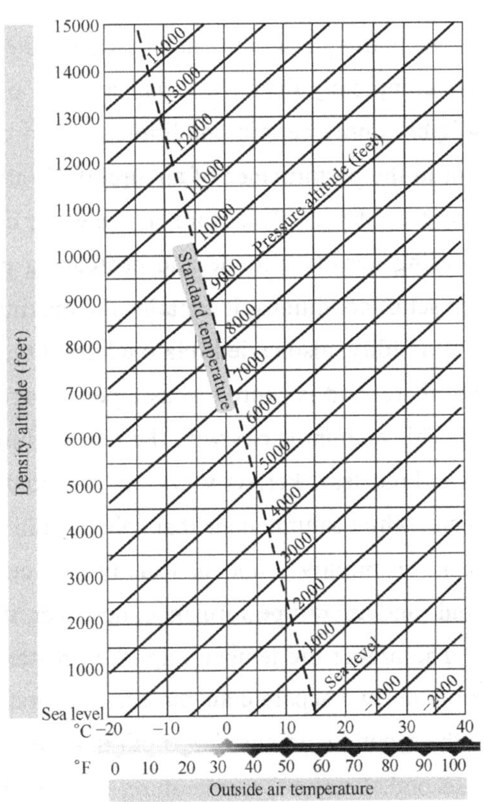

Figure 3-4 Density altitude chart

3.1.4 Effects of Pressure on Density

Since air is a gas, it can be compressed or expanded. When air is compressed, a greater amount of air can occupy a given volume. Conversely, when pressure on a given volume of air is decreased, the air expands and occupies a greater space. That is, the original column of air at a lower pressure contains a smaller mass of air. In other words, the density is decreased. In fact, density is directly proportional to pressure. If the pressure is doubled, the density is doubled, and if the pressure is lowered, so is the density. This statement is true only at a constant temperature.

3.1.5 Effects of Temperature on Density

Increasing the temperature of a substance decreases its density. Conversely, decreasing the temperature increases the density. Thus, the density of air varies inversely with temperature. This statement is true only at a constant pressure. In the atmosphere, both temperature and pressure decrease with altitude, and have conflicting effects upon density. However, the fairly rapid drop in pressure as altitude is increased usually has the dominant effect. Hence, pilots can expect the density to decrease with altitude.

3.1.6 Effects of Humidity (Moisture) on Density

The preceding paragraphs are based on the presupposition of perfectly dry air. In reality, it is never completely dry. The small amount of water vapor suspended in the atmosphere may be negligible under certain conditions, but in other conditions humidity may become an important factor in the performance of an aircraft. Water vapor is lighter than air; consequently, moist air is lighter than dry air. Therefore, as the water content of the air increases, the air becomes less dense, increasing density altitude and decreasing performance. It is lightest or least dense when, in a given set of conditions, it contains the maximum amount of water vapor.

Humidity, also called relative humidity, refers to the amount of water vapor contained in the atmosphere, and is expressed as a percentage of the maximum amount of water vapor the air can hold. This amount varies with the temperature; warm air can hold more water vapor, while colder air can hold less. Perfectly dry air that contains no water vapor has a relative humidity of zero percent, while saturated air that cannot hold any more water vapor has a relative humidity of 100 percent. Humidity alone is usually not considered an essential factor in calculating density altitude and aircraft performance; however, it does contribute.

The higher the temperature, the greater amount of water vapor that the air can hold. When comparing two separate air masses, the first warm and moist (both qualities making air lighter) and the second cold and dry (both qualities making it heavier), the first must be less dense than the second. Pressure, temperature, and humidity have a great influence on aircraft performance because of their effect upon density. There is no rule-of-thumb or chart used to compute the effects of humidity on density altitude, but it must be taken into consideration. Expect a decrease in overall performance in high humidity conditions.

3.2 Theories in the Production of Lift

3.2.1 Newton's Basic Laws of Motion

The formulation of lift has historically been the adaptation over the past few centuries of basic physical laws. These laws, although seemingly applicable to all aspects of lift, do not answer how lift is formulated. In fact, one must consider the many airfoils that are symmetrical, yet produce significant lift.

The fundamental physical laws governing the forces acting upon an aircraft in flight were adopted from postulated theories developed before any human successfully flew an aircraft. The use of these physical laws grew out of the Scientific Revolution, which began in Europe in the 1600s. Driven by the belief that the universe operated in a predictable manner open to human understanding, many philosophers, mathematicians, natural scientists, and inventors spent their lives unlocking the secrets of the universe. One of the best known was Sir Isaac Newton, who not only formulated the Law of Universal Gravitation, but also described the three basic laws of motion.

Newton's First Law: "Every object persists in its state of rest or uniform motion in a straight line unless it is compelled to change that state by forces impressed on it."

This means that nothing starts or stops moving until some outside force causes it to do so. An aircraft at rest on the ramp remains at rest unless a force strong enough to overcome its inertia is applied. Once it is moving, its inertia keeps it moving, subject to the various other forces acting on it. These forces may add to its motion, slow it down, or change its direction.

Newton's Second Law: "Force is equal to the change in momentum per change in time. For a constant mass, force equals mass times acceleration."

When a body is acted upon by a constant force, its resulting acceleration is inversely proportional to the mass of the body and is directly proportional to the applied force. This takes into account the factors involved in overcoming Newton's First Law. It covers both changes in direction and speed, including starting up from rest (positive acceleration) and coming to a stop (negative acceleration or deceleration).

Newton's Third Law: "For every action, there is an equal and opposite reaction."

In an airplane, the propeller moves and pushes back the air; consequently, the air pushes the propeller (and thus the airplane) in the opposite direction—forward. In a jet airplane, the engine pushes a blast of hot gases backward; the force of equal and opposite reaction pushes against the engine and forces the airplane forward.

3.2.2 Magnus Effect

1. Flow of Air Against a Nonrotating Cylinder

If air flows against a cylinder that is not rotating, the flow of air above and below the cylinder is identical and the forces are the same (Figure 3-5).

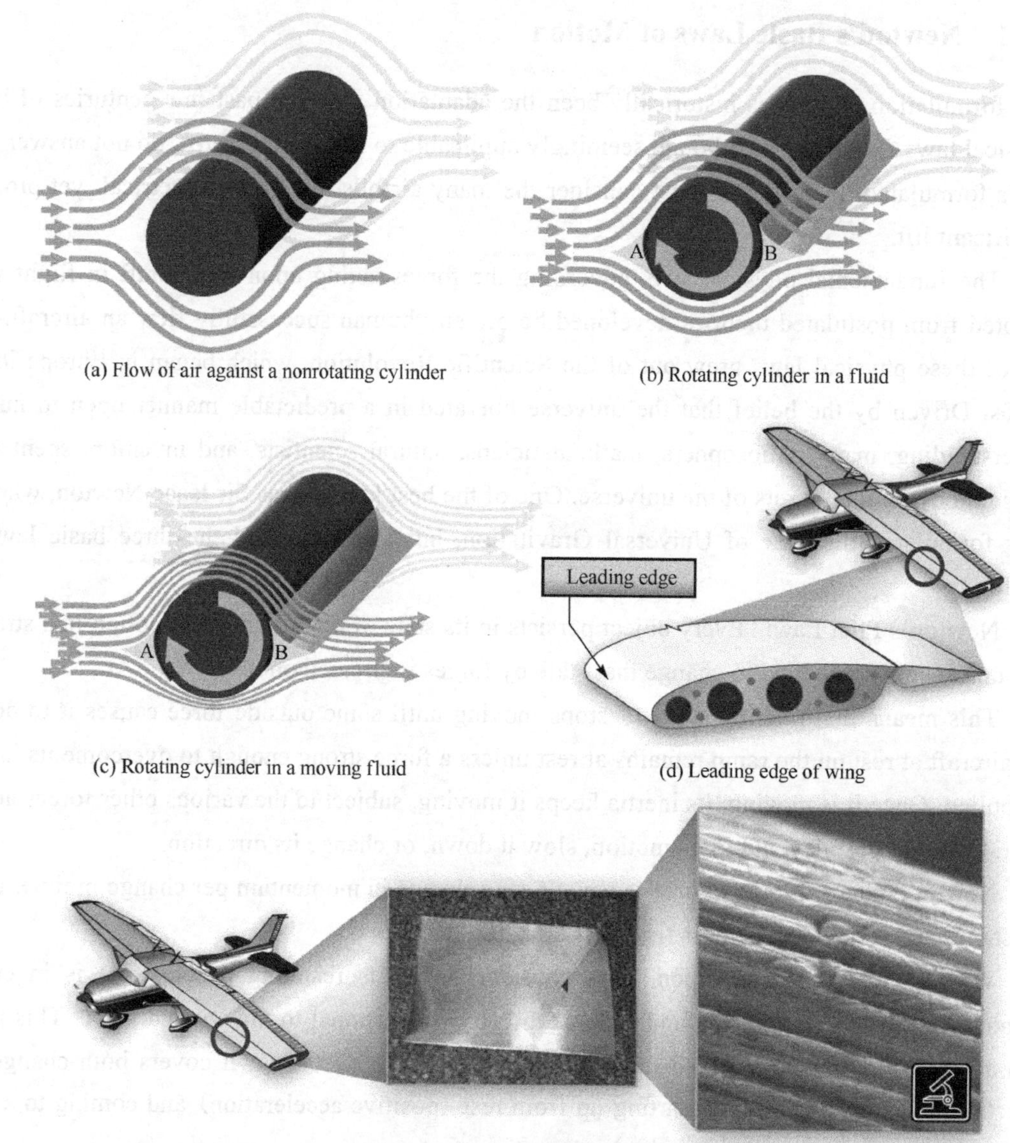

Figure 3-5 (a) illustrates uniform circulation, (b) illustrates the increased airflow over the top of a rotating cylinder, (c) The airflow speed is further increased when the rotating cylinder is in a moving stream of air, (d) is an example of typical aircraft grade aluminum used in aircraft construction to include wings and leading edges of wings as shown in (e)

2. A Rotating Cylinder in a Motionless Fluid

In Figure 3-5 (b), the cylinder is rotated clockwise and observed from the side while immersed in the fluid, the rotation of the cylinder affects the fluid surrounding the cylinder. The flow around the rotating cylinder differs from the flow around a stationary cylinder due to resistance caused by two factors: viscosity and friction.

3. Viscosity

Viscosity is the property of a fluid or semifluid that causes it to resist flowing. This resistance to flow is measurable due to the molecular tendency of fluids to adhere to each other to some extent. High-viscosity fluids resist flow; low-viscosity fluids flow easily.

Similar amounts of oil and water poured down two identical ramps demonstrate the difference in viscosity. The water seems to flow freely while the oil flows much more slowly.

Since molecular resistance to motion underlies viscosity, grease is very viscous because its molecules resist flow. Hot lava is another example of a viscous fluid. All fluids are viscous and have a resistance to flow whether this resistance is observed or not. Air is an example of a fluid whose viscosity can not be observed. Since air has viscosity properties, it will resist flow to some extent. In the case of the rotating cylinder within an immersed fluid (oil, water, or air), the fluid (no matter what it is) resists flowing over the cylinder's surface.

4. Friction

Friction is the second factor at work when a fluid flows around a rotating cylinder. Friction is the resistance one surface or object encounters when moving over another and exists between a fluid and the surface over which it flows.

If identical fluids are poured down the ramp, they flow in the same manner and at the same speed. If one ramp's surface is coated with small pebbles, the flow down the two ramps differs significantly. The rough surface ramp impedes the flow of the fluid due to resistance from the surface (friction). It is important to remember that all surfaces, no matter how smooth they appear, are not smooth and impede the flow of a fluid. Both the surface of a wing and the rotating cylinder have a certain roughness, albeit at a microscopic level, causing resistance for a fluid to flow. This reduction in velocity of the airflow about a surface is caused by skin friction or drag.

5. A Rotating Cylinder in a Moving Fluid

When the cylinder rotates in a fluid that is also moving, the result is a higher circulatory flow in the direction of the rotating cylinder. By adding fluid motion, the magnitude of the flow increases.

As shown in (Figure 3-6), at point "A", a stagnation point exists where the air stream impacts (impinges) on the front of the airfoil and splits; some air goes over, and some under. Another stagnation point exists at "B", where the two airstreams rejoin and resume at identical velocities. When viewed from the side, an up wash is created ahead of the airfoil and downwash at the rear. In the case of Figure 3-6, the highest velocity is at the top of the airfoil, and the lowest velocity at the bottom. Because these velocities are associated with an object (in this case, an airfoil) they are called local velocities as they do not exist outside the lift-producing system. This concept can be readily applied to a wing or other lifting surface.

Because there is a difference of velocity above and below the wing, the result is a higher pressure below the wing and a lower pressure above the wing. This pressure difference produces an upward force known as the Magnus Effect, the physical phenomenon whereby an object's rotation affects its path through a fluid.

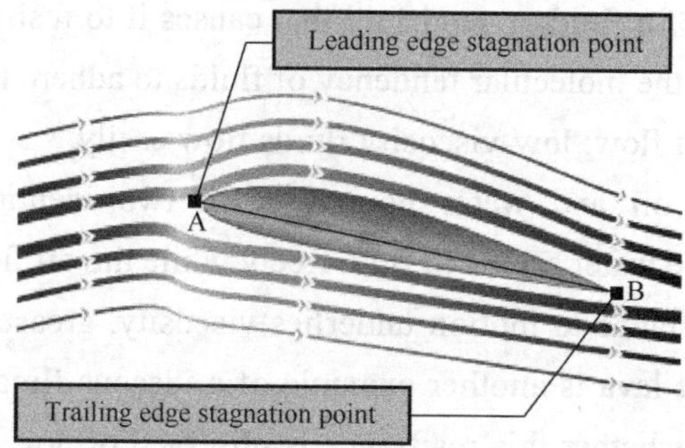

Figure 3-6 Air circulation around an airfoil occurs when the front stagnation point is below the leading edge and the aft stagnation point is beyond the trailing edge

3.2.3 Bernoulli Effect

A practical application of Bernoulli's Principle is the Venturi tube. The Venturi tube has an air inlet that narrows to a throat (constricted point) and an outlet section that increases in diameter toward the rear. The diameter of the outlet is the same as that of the inlet. At the throat, the airflow speeds up and the pressure decreases; at the outlet, the airflow slows and the pressure increases (Figure 3-7).

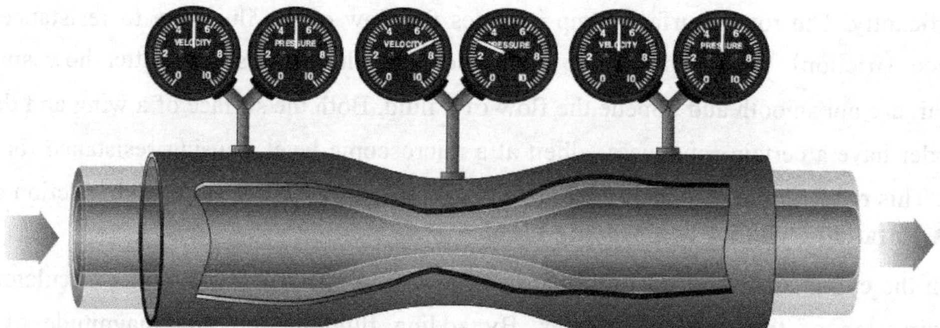

Figure 3-7 Air pressure decreases in a Venturi tube

3.3 Airfoil Design

A reference line often used in discussing the airfoil is the chord line, a straight line drawn through the profile connecting the extremities of the leading and trailing edges. The distance from this chord line to the upper and lower surfaces of the wing denotes the magnitude of the upper and lower camber at any point. Another reference line, drawn from the leading edge to the trailing edge, is the mean camber line. This mean line is equidistant at all points from the upper and lower surfaces (Figure 3-8).

An airfoil is constructed in such a way that its shape takes advantage of the air's response to certain physical laws. This develops two actions from the air mass: a positive pressure lifting action from the air mass below the wing, and a negative pressure lifting action from lowered pressure above the wing.

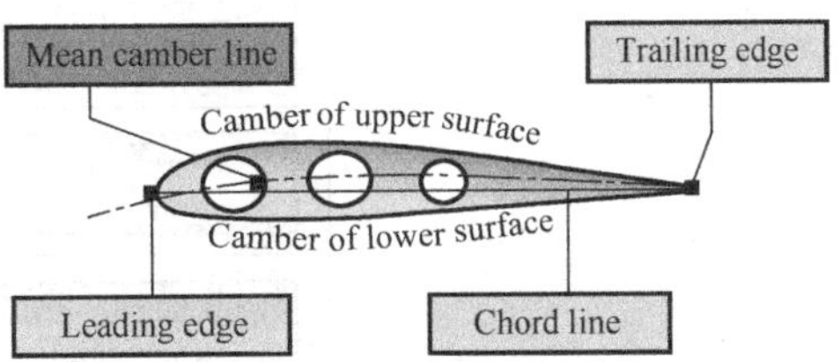

Figure 3-8 Typical airfoil section

As the air stream strikes the relatively flat lower surface of a wing or rotor blade when inclined at a small angle to its direction of motion, the air is forced to rebound downward, causing an upward reaction in positive lift. At the same time, the air stream striking the upper curved section of the leading edge is deflected upward. An airfoil is shaped to cause an action on the air, and forces air downward, which provides an equal reaction from the air, forcing the airfoil upward. If a wing is constructed in such form that it causes a lift force greater than the weight of the aircraft, the aircraft will fly.

If all the lift required were obtained merely from the deflection of air by the lower surface of the wing, an aircraft would only need a flat wing like a kite. However, the balance of the lift needed to support the aircraft comes from the flow of air above the wing. Herein lies the key to flight.

It is neither accurate nor useful to assign specific values to the percentage of lift generated by the upper surface of an airfoil versus that generated by the lower surface. These are not constant values and vary, not only with flight conditions, but also with different wing designs.

Different airfoils have different flight characteristics. Many thousands of airfoils have been tested in wind tunnels and in actual flight, but no one airfoil has been found that satisfies every flight requirement. The weight, speed, and purpose of each aircraft dictate the shape of its airfoil. The most efficient airfoil for producing the greatest lift is one that has a concave, or "scooped out" lower surface. As a fixed design, this type of airfoil sacrifices too much speed while producing lift and is not suitable for high-speed flight. Advancements in engineering have made it possible for today's high-speed jets to take advantage of the concave airfoil's high lift characteristics. Leading edge (Kreuger) flaps and trailing edge (Fowler) flaps, when extended from the basic wing structure, literally change the airfoil shape into the classic concave form, thereby generating much greater lift during slow flight conditions.

On the other hand, an airfoil that is perfectly streamlined and offers little wind resistance sometimes does not have enough lifting power to take the airplane off the ground. Thus, modern airplanes have airfoils that strike a medium between extremes in design. The shape varies according to the needs of the airplane for which it is designed. Figure 3-9 shows some of the more common airfoil sections.

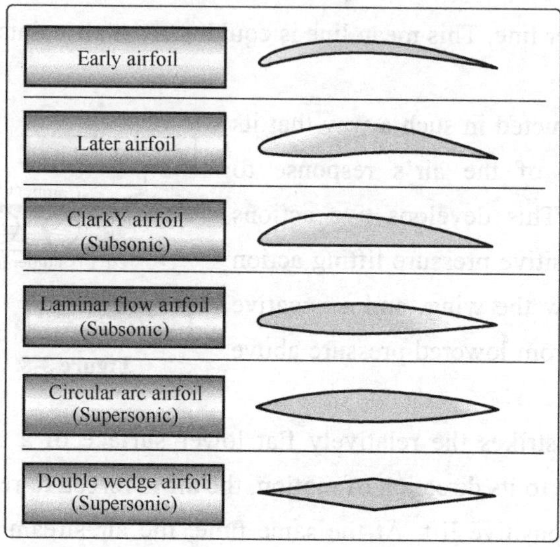

Figure 3-9 Airfoil designs

3.3.1 Low Pressure Above

In a wind tunnel or in flight, an airfoil is simply a streamlined object inserted into a moving stream of air. If the airfoil profile were in the shape of a teardrop, the speed and the pressure changes of the air passing over the top and bottom would be the same on both sides. But if the teardrop shaped airfoil were cut in half lengthwise, a form resembling the basic airfoil (wing) section would result. If the airfoil were then inclined to make the airflow strike it at an angle (angle of attack), the air moving over the upper surface would be forced to move faster than the air moving along the bottom of the airfoil. This increased velocity reduces the pressure above the airfoil.

3.3.2 High Pressure Below

A certain amount of lift is generated by pressure conditions underneath the airfoil. Because of the manner in which air flows underneath the airfoil, a positive pressure results, particularly at higher angles of attack. But there is another aspect to this airflow that must be considered. At a point close to the leading edge, the airflow is virtually stopped (stagnation point) and then gradually increases speed. At some point near the trailing edge, it again reaches a velocity equal to that on the upper surface. In conformance with Bernoulli's principle, where the airflow was slowed beneath the airfoil, a positive upward pressure was created i.e., as the fluid speed decreases, the pressure must increase. Since the pressure differential between the upper and lower surface of the airfoil increases, total lift increases. Both Bernoulli's Principle and Newton's Laws are in operation whenever lift is being generated by an airfoil.

3.3.3 Pressure Distribution

From experiments conducted on wind tunnel models and on full size airplanes, it has been determined that as air flows along the surface of a wing at different angles of attack, there are regions along the surface where the pressure is negative, or less than atmosphere, and regions where the pressure is positive, or greater than atmosphere. This negative pressure on the upper surface creates a relatively larger force on the wing that is caused by the positive pressure resulting from the air striking the lower wing surface. Figure 3-10 shows the pressure distribution along an airfoil at three different angles of attack. The average of the pressure variation for any given angle of attack is referred to as the center of pressure (CP). Aerodynamic force acts through this CP. At high angles of attack, the CP moves forward, while at low angles of attack the CP moves aft. In the design of wing structures, this CP travel is very important, since it affects the position of the air loads imposed on the wing structure in both low and high AOA conditions. An airplane's aerodynamic balance and controllability are governed by changes in the CP.

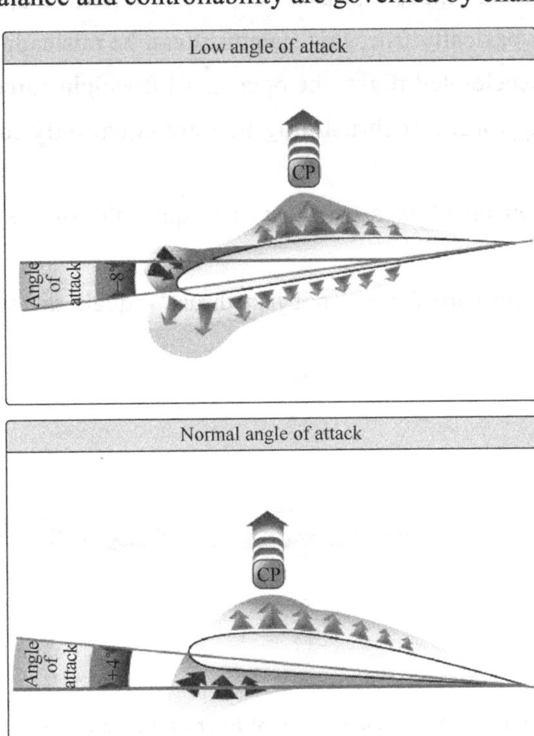

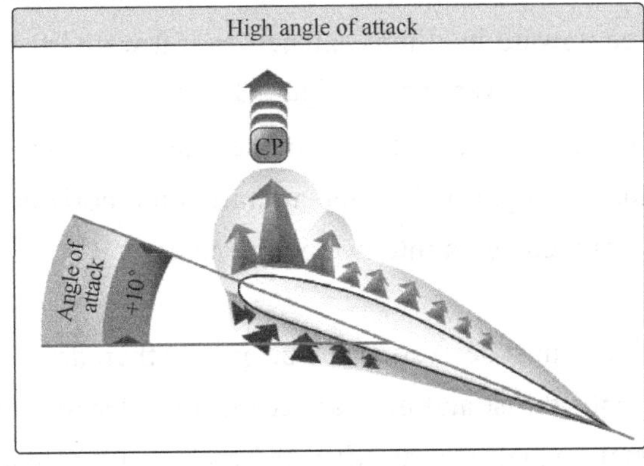

Figure 3-10 Pressure distribution on an airfoil and CP changes with AOA

3.4 Principles of Flight

In steady flight, the sum of the opposing forces is always zero. There can be no unbalanced forces in steady, straight flight based upon Newton's Second Law. Newton's Third Law states that for every action or force there is an equal, but opposite, reaction. This is true whether an airplane is flying horizontally, climbing or descending.

It does not mean the four forces are equal. It means the opposing forces are equal, and thereby cancel the effects of each other. In Figure 3-11 the force vectors of thrust, drag, lift and weight appear to be equal in value. The usual explanation states that thrust equals drag and lift equals weight. Although basically true, this statement can be misleading. It should be understood that in straight level, unaccelerated flight, the opposing lift/weight forces are equal. They are also greater than the opposing forces of thrust/drag that are equal only to each other. Therefore, in steady flight:

(1) The sum of all upward forces (not just lift) equals the sum of all downward forces (not just weight).

(2) The sum of all forward forces (not just thrust) equals the sum of all backward forces (not just drag).

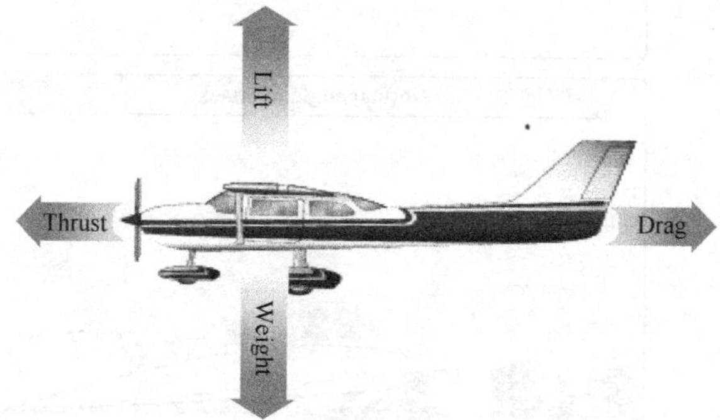

Figure 3-11 Relationship of forces acting on an airplane

This refinement of the old "thrust equals drag; lift equals weight" formula explains that a portion of thrust is directed upward in climbs and acts as if it were lift while a portion of weight is directed backward and acts as if it were drag (Figure 3-12).

In glides, a portion of the weight vector is directed forward, and, therefore, acts as thrust. In other words, any time the flight path of the aircraft is not horizontal, lift, weight, thrust, and drag vectors must each be broken down into two components.

1. Thrust

For an aircraft to move, thrust must be exerted and be greater than drag. The aircraft will continue to move and gain speed until thrust and drag are equal. In order to maintain a constant airspeed, thrust and drag must remain equal, just as lift and weight must be equal to maintain a constant altitude. If in level flight, the engine power is reduced, the thrust is lessened, and the aircraft slows down. As long as the thrust is less than the drag, the aircraft continues to decelerate until its airspeed is insufficient to maitain flight.

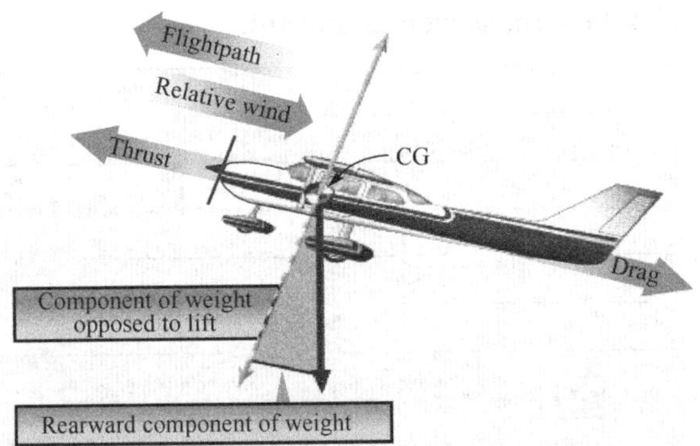

Figure 3-12 Force vectors during a stabilized climb

Likewise, if the engine power is increased, thrust becomes greater than drag and the airspeed increases. As long as the thrust continues to be greater than the drag, the aircraft continues to accelerate. When drag equals thrust, the aircraft flies at a constant airspeed.

2. Drag

Drag is the force that resists movement of an aircraft through the air. There are two basic types: parasite drag and induced drag. The first is called parasite because it in no way functions to aid flight, while the second, induced drag, is a result of an airfoil developing lift.

3. Weight

Gravity is the pulling force that tends to draw all bodies to the center of the earth. The CG may be considered as a point at which all the weight of the aircraft is concentrated. If the aircraft were supported at its exact CG, it would balance in any attitude. It should be noted that CG is of major importance in an aircraft, for its position has a great bearing upon stability.

The location of the CG is determined by the general design of each particular aircraft. The designers determine how far the center of pressure (CP) will travel. They then fix the CG forward of the center of pressure for the corresponding flight speed in order to provide an adequate restoring moment to retain flight equilibrium.

4. Lift

The pilot can control the lift. Any time the control yoke or stick is moved fore or aft, the AOA is changed. As the AOA increases, lift increases (all other factors being equal). When the aircraft reaches the maximum AOA, lift begins to diminish rapidly. This is the stalling AOA, known as C_L-MAX critical AOA. Examine Figure 3-13, noting how the C_L increases until the critical AOA is reached, then decreases rapidly with any further increase in the AOA.

5. Forces in Turns

If an aircraft was viewed in straight-and-level flight from the front Figure 3-14, and if the forces acting on the aircraft could be seen, lift and weight would be apparent: two forces. If the aircraft was in a bank it would be apparent that lift did not act directly opposite to the weight, rather it now acts in the direction of the bank. A basic truth about turns: when the aircraft banks, lift acts inward toward the center of the turn, as well as upward.

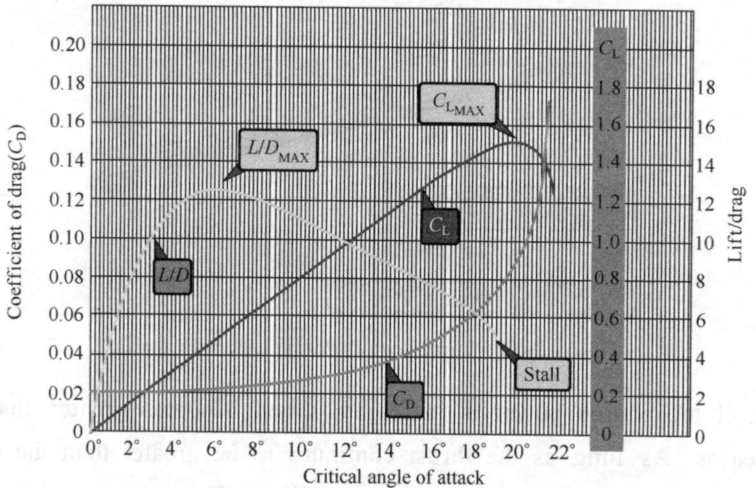

Figure 3-13 Lift coefficients at various angles of attack

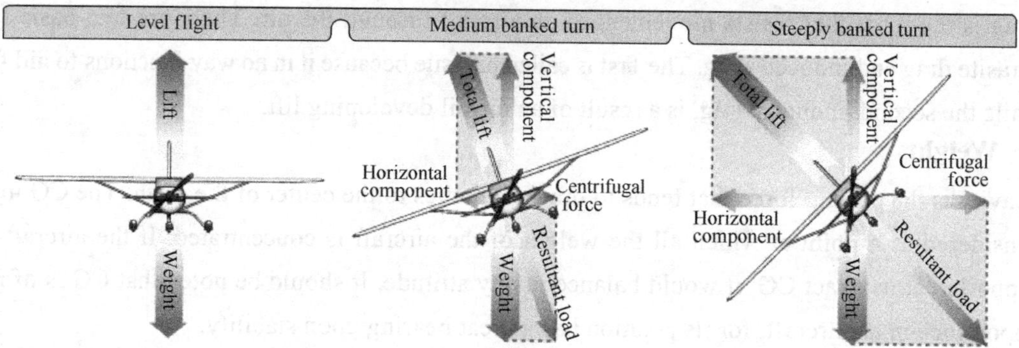

Figure 3-14 Forces during normal coordinated turn

Newton's First Law of Motion, the Law of Inertia, states that an object at rest or moving in a straight line remains at rest or continues to move in a straight line until acted on by some other force. An aircraft, like any moving object, requires a sideward force to make it turn. In a normal turn, this force is supplied by banking the aircraft so that lift is exerted inward, as well as upward. The force of lift during a turn is separated into two components at right angles to each other. One component, which acts vertically and opposite to the weight (gravity), is called the "vertical component of lift". The other, which acts horizontally toward the center of the turn, is called the "horizontal component of lift", or centripetal force. The horizontal component of lift is the force that pulls the aircraft from a straight flight path to make it turn. Centrifugal force is the "equal and opposite reaction" of the aircraft to the change in direction and acts equal and opposite to the horizontal component of lift. This explains why, in a correctly executed turn, the force that turns the aircraft is not supplied by the rudder. The rudder is used to correct any deviation between the straight track of the nose and tail of the aircraft. A good turn is one in which the nose and tail of the aircraft track along the same path. If no rudder is used in a turn, the nose of the aircraft yaws to the outside of the turn. The rudder is used to bring the nose back in line with the relative wind.

In order for an aircraft to turn, it must be banked. If it is not banked, there is no force available to cause it to deviate from a straight flight path. Conversely, when an aircraft is banked, it turns, provided it is not slipping to the inside of the turn. Good directional control is based on the fact that the aircraft attempts to turn whenever it is banked. Pilots should keep this fact in mind when attempting to hold the aircraft in straight-and-level flight. Merely banking the aircraft into a turn produces no change in the total amount of lift developed. Since the lift during the bank is divided into vertical and horizontal components, the amount of lift opposing gravity and supporting the aircraft's weight is reduced. Consequently, the aircraft loses altitude unless additional lift is created. This is done by increasing the AOA until the vertical component of lift is again equal to the weight. Since the vertical component of lift decreases as the bank angle increases, the AOA must be progressively increased to produce sufficient vertical lift to support the aircraft's weight. An important fact for pilots to remember when making constant altitude turns is that the vertical component of lift must be equal to the weight to maintain altitude.

3.5 Aircraft Axes and Stability

The axes of an aircraft are three imaginary lines that pass through an aircraft's CG. The axes can be considered as imaginary axles around which the aircraft turns. The three axes pass through the CG at 90° angles to each other. The axis from nose to tail is the longitudinal axis, the axis that passes from wingtip to wingtip is the lateral axis, and the axis that passes vertically through the CG is the vertical axis. Whenever an aircraft changes its flight attitude or position in flight, it rotates about one or more of the three axes (Figure 3-15).

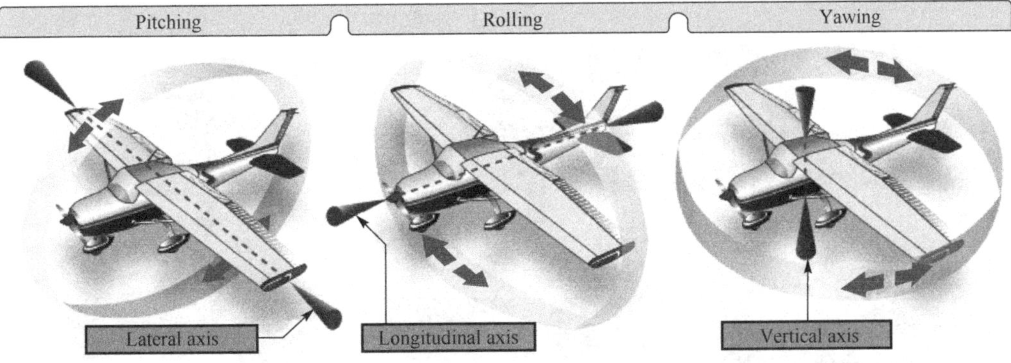

Figure 3-15 Axes of an airplane

The aircraft's motion about its longitudinal axis resembles the roll of a ship from side to side. In fact, the names used to describe the motion about an aircraft's three axes were originally nautical terms. They have been adapted to aeronautical terminology due to the similarity of motion of aircraft and seagoing ships. The motion about the aircraft's longitudinal axis is "roll", the motion about its lateral axis is "pitch", and the motion about its vertical axis is "yaw". Yaw is the horizontal (left and right) movement of the aircraft's nose.

Stability is the inherent quality of an aircraft to correct for conditions that may disturb its equilibrium, and to return to or to continue on the original flight path. It is primarily an aircraft design characteristic. The flight paths and attitudes an aircraft flies are limited by the aerodynamic characteristics of the aircraft, its propulsion system, and its structural strength. These limitations indicate the maximum performance and maneuverability of the aircraft. If the aircraft is to provide maximum utility, it must be safely controllable to the full extent of these limits without exceeding the pilot's strength or requiring exceptional flying ability. If an aircraft is to fly straight and steady along any arbitrary flight path, the forces acting on it must be in static equilibrium. The reaction of any body when its equilibrium is disturbed is referred to as stability. The two types of stability are static and dynamic.

1. Static Stability

Static stability refers to the initial tendency, or direction of movement, back to equilibrium. In aviation, it refers to the aircraft's initial response when disturbed from a given AOA, slip, or bank (Figure 3-16).

(1) Positive static stability—the initial tendency of the aircraft to return to the original state of equilibrium after being disturbed.

(2) Neutral static stability—the initial tendency of the aircraft to remain in a new condition after its equilibrium has been disturbed.

(3) Negative static stability—the initial tendency of the aircraft to continue away from the original state of equilibrium after being disturbed.

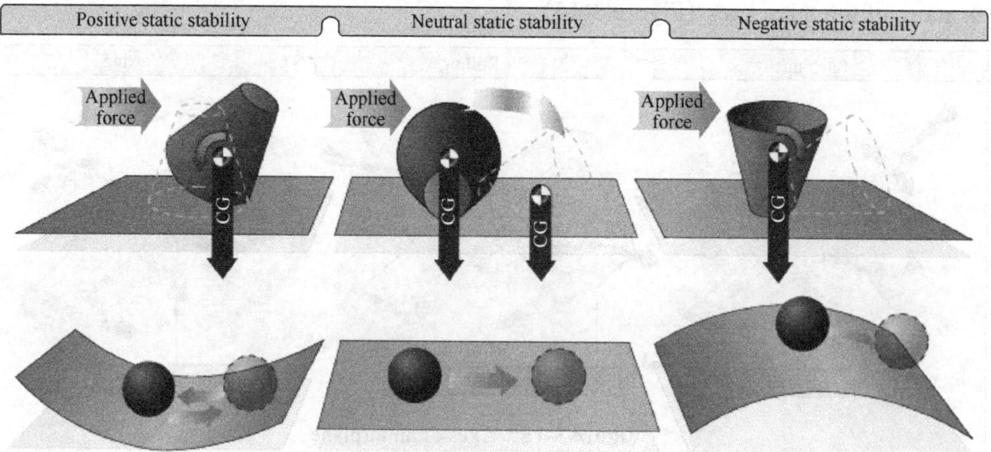

Figure 3-16 Types of static stability

2. Dynamic Stability

Static stability has been defined as the initial tendency to return to equilibrium that the aircraft displays after being disturbed from its trimmed condition. Occasionally, the initial tendency is different or opposite from the overall tendency, so a distinction must be made between the two. Dynamic stability refers to the aircraft response over time when disturbed from a given AOA, slip, or bank. This type of stability also has three subtypes:

(1) Positive dynamic stability—over time, the motion of the displaced object decreases in amplitude and, because it is positive, the object displaced returns toward the equilibrium state.

(2) Neutral dynamic stability—once displaced, the displaced object neither decreases nor increases in amplitude. A worn automobile shock absorber exhibits this tendency.

(3) Negative dynamic stability—over time, the motion of the displaced object increases and becomes more divergent.

Stability in an aircraft affects two areas significantly:

(1) Maneuverability—the quality of an aircraft that permits it to be maneuvered easily and to withstand the stresses imposed by maneuvers. It is governed by the aircraft's weight, inertia, size and location of flight controls, structural strength, and powerplant. It too is an aircraft design characteristic.

(2) Controllability—the capability of an aircraft to respond to the pilot's control, especially with regard to flight path and attitude. It is the quality of the aircraft's response to the pilot's control application when the pilot maneuvers the aircraft, regardless of its stability characteristics.

3.6 Flight Control

Aircraft flight control systems consist of primary and secondary systems. The ailerons, elevator (or stabilator) and rudder constitute the primary control system and are required to control an aircraft safely during flight. Wing flaps, leading edge devices, spoilers, and trim systems constitute the secondary control system and improve the performance characteristics of the airplane or relieve the pilot of excessive control forces.

3.6.1 Primary Flight Control

Aircraft control systems are carefully designed to provide adequate responsiveness to control inputs while allowing a natural feel. At low airspeeds, the controls usually feel soft and sluggish, and the aircraft responds slowly to control applications. At higher airspeeds, the controls become increasingly firm and aircraft response is more rapid.

Movement of any of the three primary flight control surfaces (ailerons, elevator or stabilator, or rudder), changes the airflow and pressure distribution over and around the airfoil. These changes affect the lift and drag produced by the airfoil/control surface combination and allow a pilot to control the aircraft about its three axes of rotation.

Design features limit the amount of deflection of flight control surfaces. For example, control-stop mechanisms may be incorporated into the flight control linkages, or movement of the control column and/or rudder pedals may be limited. The purpose of these design limits is to prevent the pilot from inadvertently overcontrolling and overstressing the aircraft during normal maneuvers. A properly designed airplane is stable and easily controlled during normal maneuvering. Control surface inputs cause movement about the three axes of rotation. The types of stability an airplane exhibits also relate to the three axes of rotation (Figure 3-17).

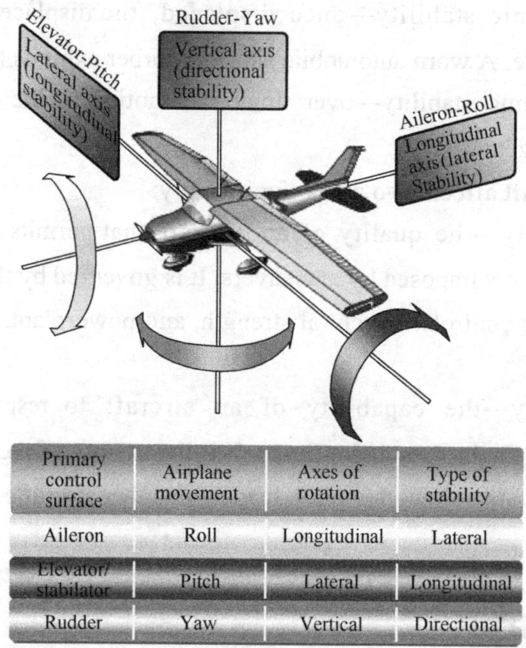

Primary control surface	Airplane movement	Axes of rotation	Type of stability
Aileron	Roll	Longitudinal	Lateral
Elevator/ stabilator	Pitch	Lateral	Longitudinal
Rudder	Yaw	Vertical	Directional

Figure 3-17 Airplane controls, movement, axes of rotation, and type of stability

1. Ailerons

Ailerons control roll about the longitudinal axis. The ailerons are attached to the outboard trailing edge of each wing and move in the opposite direction from each other. Ailerons are connected by cables, bell cranks, pulleys and/or push-pull tubes to a control wheel or control stick.

Moving the control wheel or control stick to the right causes the right aileron to deflect upward and the left aileron to deflect downward. The upward deflection of the right aileron decreases the camber resulting in decreased lift on the right wing. The corresponding downward deflection of the left aileron increases the camber resulting in increased lift on the left wing. Thus, the increased lift on the left wing and the decreased lift on the right wing cause the airplane to roll to the right.

2. Adverse Yaw

Since the downward deflected aileron produces more lift as evidenced by the wing rising, it also produces more drag. This added drag causes the wing to slow down slightly. This results in the aircraft yawing toward the wing which had experienced an increase in lift (and drag). From the pilot's perspective, the yaw is opposite the direction of the bank. The adverse yaw is a result of differential drag and the slight difference in the velocity of the left and right wings (Figure 3-18).

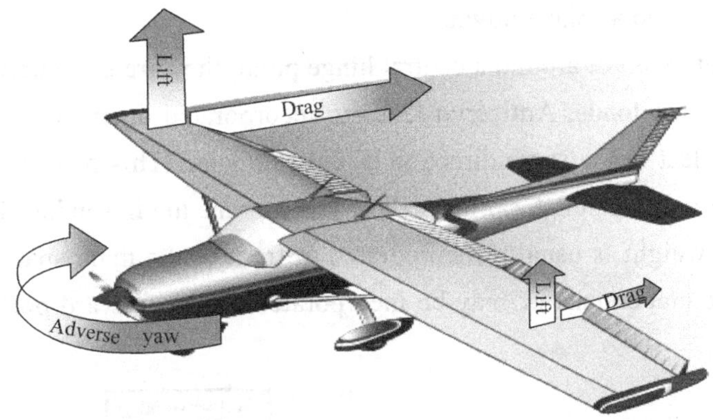

Figure 3-18 Adverse yaw is caused by higher drag on the outside wing, which is producing more lift

3. Elevator

The elevator controls pitch about the lateral axis. Like the ailerons on small aircraft, the elevator is connected to the control column in the flight deck by a series of mechanical linkages. Aft movement of the control column deflects the trailing edge of the elevator surface up. This is usually referred to as up "elevator" (Figure 3-19).

The up-elevator position decreases the camber of the elevator and creates a downward aerodynamic force, which is greater than the normal tail-down force that exists in straight-and-level flight. The overall effect causes the tail of the aircraft to move down and the nose to pitch up. The pitching moment occurs about the center of gravity (CG). The strength of the pitching moment is determined by the distance between the CG and the horizontal tail surface, as well as by the aerodynamic effectiveness of the horizontal tail surface. Moving the control column forward has the opposite effect. In this case, elevator camber increases, creating more lift (less tail-down force) on the horizontal stabilizer/elevator. This moves the tail upward and pitches the nose down. Again, the pitching moment occurs about the CG.

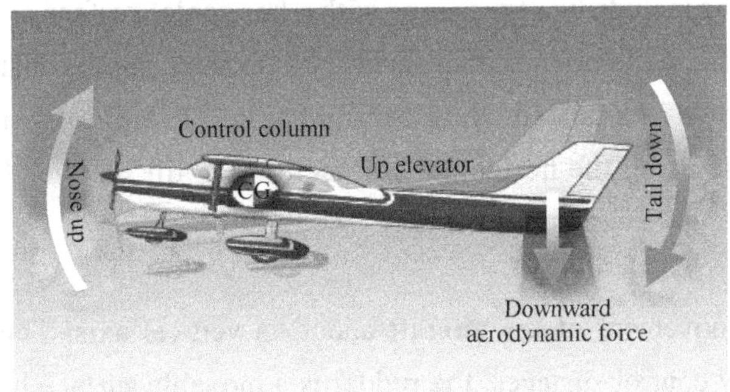

Figure 3-19 The elevator is the primary control for changing the pitch attitude of an airplane

4. Stabilator

A stabilator is essentially a one-piece horizontal stabilizer that pivots from a central hinge point. When the control column is pulled back, it raises the stabilator's trailing edge, pulling the airplane's nose up. Pushing the control column forward lowers the trailing edge of the stabilator and pitches the nose of the airplane down.

Because stabilators pivot around a central hinge point, they are extremely sensitive to control inputs and aerodynamic loads. Antiservo tabs are incorporated on the trailing edge to decrease sensitivity. They deflect in the same direction as the stabilator. This results in an increase in the force required to move the stabilator, thus making it less prone to pilot-induced overcontrolling. In addition, a balance weight is usually incorporated in front of the main spar. The balance weight may project into the empennage or may be incorporated on the forward portion of the stabilator tips (Figure 3-20).

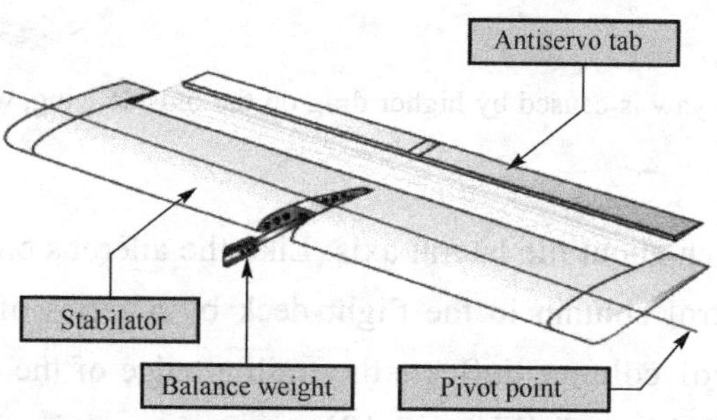

Figure 3-20 The stabilator

5. Canard

The canard design utilizes the concept of two lifting surfaces, the canard functioning as a horizontal stabilizer located in front of the main wings. In effect, the canard is an airfoil similar to the horizontal surface on a conventional aft-tail design. The difference is that the canard actually creates lift and holds the nose up, as opposed to the aft-tail design which exerts downward force on the tail to prevent the nose from rotating downward.

The canard design dates back to the pioneer days of aviation, most notably used on the Wright Flyer. Recently, the canard configuration has regained popularity and is appearing on newer aircraft. Canard designs include two types–one with a horizontal surface of about the same size as a normal aft-tail design, and the other with a surface of the same approximate size and airfoil of the aft-mounted wing known as a tandem wing configuration. Theoretically, the canard is considered more efficient because using the horizontal surface to help lift the weight of the aircraft should result in less drag for a given amount of lift.

6. Rudder

The rudder controls movement of the aircraft about its vertical axis. This motion is called yaw. Like the other primary control surfaces, the rudder is a movable surface hinged to a fixed surface, in this case to the vertical stabilizer, or fin. Moving the left or right rudder pedal controls the rudder.When the rudder is deflected into the airflow, a horizontal force is exerted in the opposite direction (Figure 3-21). By pushing the left pedal, the rudder moves left. This alters the airflow around the vertical stabilizer/rudder and creates a sideward lift that moves the tail to the right and yaws the nose of the airplane to the left. Rudder effectiveness increases with speed; therefore, large deflections at low speeds and small deflections at high speeds may be required to provide the desired reaction. In propeller-driven aircraft, any slipstream flowing over the rudder increases its effectiveness.

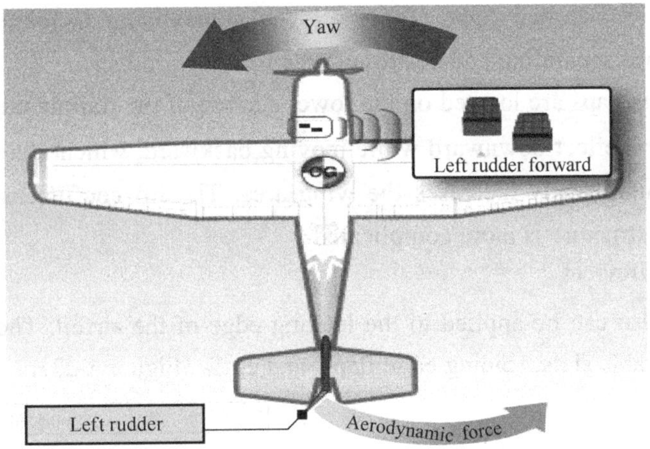

Figure 3-21 The effect of left rudder pressure

3.6.2 Secondary Flight Control

1. Flaps

Flaps are the most common high-lift devices used on aircraft. These surfaces, which are attached to the trailing edge of the wing, increase both lift and induced drag for any given AOA. Flaps allow a compromise between high cruising speed and low landing speed, because they may be extended when needed, and retracted into the wing's structure when not needed. There are four common types of flaps: plain, split, slotted, and Fowler flaps (Figure 3-22).

(1) The plain flap is the simplest of the four types. It increases the airfoil camber, resulting in a significant increase in the coefficient of lift (CL) at a given AOA. At the same time, it greatly increases drag and moves the center of pressure (CP) aft on the airfoil, resulting in a nose-down pitching moment.

(2) The split flap is deflected from the lower surface of the airfoil and produces a slightly greater increase in lift than the plain flap. More drag is created because of the turbulent air pattern produced behind the airfoil. When fully extended, both plain and split flaps produce high drag with little additional lift.

(3) The most popular flap on aircraft today is the slotted flap. Variations of this design are used for small aircraft, as well as for large ones. Slotted flaps increase the lift coefficient significantly more than plain or split flaps. On small aircraft, the hinge is located below the lower surface of the flap, and when the flap is lowered, a duct forms between the flap well in the wing

and the leading edge of the flap. When the slotted flap is lowered, high energy air from the lower surface is ducted to the flap's upper surface. The high energy air from the slot accelerates the upper surface boundary layer and delays airflow separation, providing a higher C_L. Thus, the slotted flap produces much greater increases in maximum coefficient of lift (C_L-MAX) than the plain or split flap. While there are many types of slotted flaps, large aircraft often have double-and even triple-slotted flaps. These allow the maximum increase in drag without the airflow over the flaps separating and destroying the lift they produce.

(4) The fowler flaps are located on the lower surface of the trailing edge of the wing. When it is lowered, it can deflect downward while moving backward, which increases the curvature of the wing section and greatly increases the wing area. The lift coefficient can be increased by 85%–95%, but the structure is more complicated.

2. Leading Edge Devices

High-lift devices also can be applied to the leading edge of the airfoil. The most common types are fixed slots, movable slats, leading edge flaps, and cuffs (Figure 3-23).

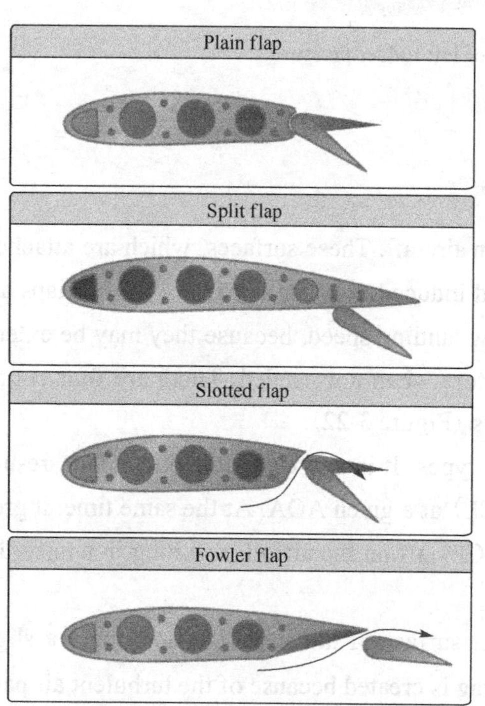

Figure 3-22 Four common types of flaps

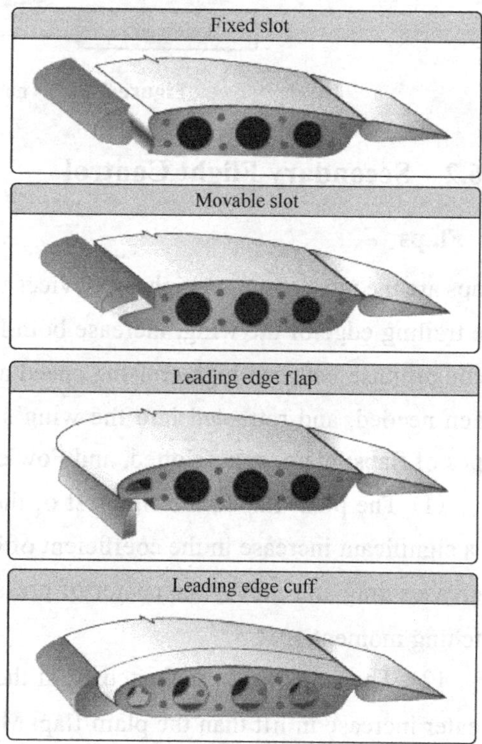

Figure 3-23 Leading edge high lift devices

Fixed slots direct airflow to the upper wing surface and delay airflow separation at higher angles of attack. The slot does not increase the wing camber, but allows a higher maximum CL because the stall is delayed until the wing reaches a greater AOA.

3. Spoilers

Found on many gliders and some aircraft, high drag devices called spoilers are deployed from the wings to spoil the smooth airflow, reducing lift and increasing drag. On gliders, spoilers are most often used to control rate of descent for accurate landings. On other aircraft, spoilers are often used for roll control, an advantage of which is the elimination of adverse yaw. To turn right, for example, the spoiler on the right wing is raised, destroying some of the lift and creating more drag on the right. The right wing drops and the aircraft banks and yaws to the right. Deploying spoilers on both wings at the same time allows the aircraft to descend without gaining speed. Spoilers are also deployed to help reduce ground roll after landing. By destroying lift, they transfer weight to the wheels, improving braking effectiveness (Figure 3-24).

Figure 3-24 Spoilers reduce lift and increase drag during descent

第3章 飞行知识

本章介绍了飞机飞行时的相关知识，包括飞行中作用于飞机上力的基本物理定律，以及这些自然定律和力对飞机性能特性的影响，飞行原理和飞机控制。要控制飞机、直升机、滑翔机或热气球，飞行员必须理解其中涉及的原理，学会利用和回避这些自然力。

3.1 大气结构

大气是包围着地球的空气层，并且依附在地球的表面。它跟海洋和陆地一样是地球的重要组成部分。然而，大气不同于土壤和水，因为它是气体的混合物，有质量、重量和不确定的形状。

空气和其他任何流体一样，可以流动，当受到瞬间的压力而由于缺少很强的分子凝聚力时，它会改变形状。例如，气体可以完全充满它所处的任何容器，通过膨胀或者收缩来改变形状，从而充满容器。大气由78%的氮气、21%的氧气和1%的其他气体(如氩气或氦气)组成。大部分氧气存在于35000ft(英尺)(1ft = 0.3048m)高度以下。

3.1.1 大气压力

尽管有很多种压力，但是飞行员主要考虑大气压力。它是天气变化的基本因素之一。大气帮助抬升飞机，还能驱动飞机上一些重要的飞行仪表。这些仪表包括高度计、空速表、垂直速率表和进气压力表(或歧管压力表)。

虽然空气很轻，但是它有质量而且受重力影响。因此，和其他任何物质一样，它有重量，而且由于它有重量，因此它就有了力。因为它是流体物质，这个力在所有方向上是相等的，它对空气中物体上的作用称为压力。在海平面标准条件下，大气重量所施加的平均压力大约为14.7磅/平方英寸($1in^2=6.4516\times10^{-4}m^2$)。空气密度对飞机的性能有重要的影响。当空气密度变小时，它会降低：

(1)功率，因为发动机吸入的空气变少。

(2)推力，因为螺旋桨在稀薄空气中效率更低。

(3)升力，因为稀薄空气对机翼施加的力更少。

大气压力随时间和地点而变化。由于大气压总是变化的，我们确定了一个标准的参考气压。在海平面的标准大气的表面温度为59℉或者15℃，且表面气压为29.92in Hg(英寸汞柱)($1in\ Hg=3.386\times10^3Pa$)或者1013.2mbar(毫巴)(1mbar=100Pa)，如图3-1所示。

标准条件下，高度每上升1000ft，温度下降约3.5℉或者2℃，上限高度达到36000ft。在这点之上，温度被认为是恒定的，直到80000ft。标准条件下，高度每上升1000ft，压力下降约1in Hg，直到10000ft高度，如图3-2所示。

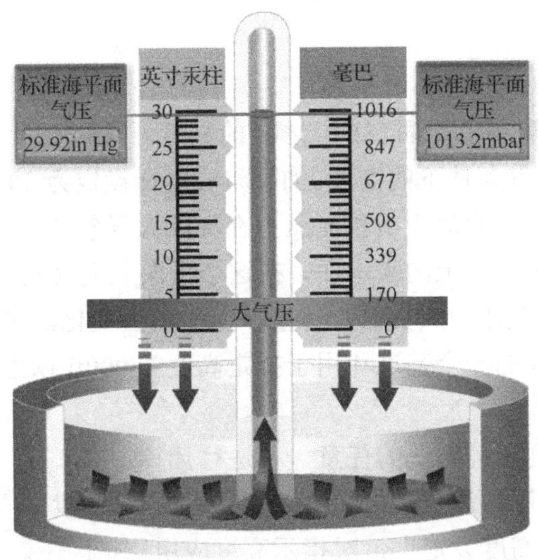

图 3-1 标准海平面气压

高度/ft	压强 /in Hg	温度	
		/℃	/°F
0	29.92	15.0	59.0
1000	28.86	13.0	55.4
2000	27.82	11.0	51.8
3000	26.82	9.1	48.4
4000	25.84	7.1	44.8
5000	24.89	5.1	41.2
6000	23.98	3.1	37.6
7000	23.09	1.1	34.0
8000	22.22	−0.9	30.4
9000	21.38	−2.8	27
10000	20.57	−4.8	23.4
11000	19.79	−6.8	19.8
12000	19.02	−8.8	16.2
13000	18.29	−10.8	12.6
14000	17.57	−12.7	9.1
15000	16.88	−14.7	5.5
16000	16.21	−16.7	1.9
17000	15.56	−18.7	−1.7
18000	14.94	−20.7	−5.3
19000	14.33	−22.6	−8.7
20000	13.74	−24.6	−12.3

图 3-2 标准大气性质

国际民用航空组织(ICAO)把这个确立为世界标准，通常称为国际标准大气(ISA)或者ICAO标准大气。任何不同于标准下降率的温度或者压力被认为是非标准温度或非标准压力。制造商的性能图表上会提供非标准温度和压力的调整办法。

因为所有飞机性能是相对于标准大气来比较和计算的，所以所有飞机仪表都按标准大气条件校准，如果实际运行条件不符合标准大气，必须对仪表的使用和飞机的性能做出某种修正。为了正确地说明标准大气，就必须定义一些相关的术语①。

① 国际标准大气(ISA)也称为标准大气，是不同高度上大气空气压力、温度和密度的代表性参考模型。在海平面，国际标准大气的温度为59℉或15℃，压力为29.92in Hg或者1013.2mbar。

3.1.2 压力高度

压力高度是位于标准参考平面(SDP)之上的高度。飞机高度计是一个重要的灵敏气压计,用以校准指示标准大气条件下的高度。如果将高度计设定为 29.92in Hg 的标准参考平面,高度计指示的是压力高度,它对应所检测压力在标准大气条件下的高度。

标准参考平面是一个理论上的水平面,在这个平面上大气压强 29.92in Hg,大气重量为 14.7 磅/平方英寸。当大气压力改变时,标准参考平面会变化,可能低于、等于或者高于海平面。作为计算飞机性能的一个基准和用于指定高度 18000ft 以上飞机运行的高度层,压力高度很重要。

压力高度可以用下列两种方法的任意一种进行计算:

(1) 设定高度计的气压计读数到 29.92,然后读出指示高度。

(2) 对应于报告的“高度设定”,对指示高度应用修正因子。

3.1.3 密度高度

和非标准大气条件下的空气动力学性能有关的术语是密度高度,它对应特定空气密度时的标准大气条件下的高度。

密度高度是经非标准温度修正后的压力高度。当空气的密度增加(较低的密度高度)时,飞机性能提高;相反,随着空气密度降低(较高的密度高度)时,飞机性能下降。密度高度用于计算飞机性能。在标准大气条件下,大气中每个高度上的空气都有特定的密度,压力高度和密度高度表示的高度相同。密度高度是标准大气条件下给定密度位置在海平面上的垂直距离。

计算密度高度必须考虑压力(压力高度)和温度。因为任何高度上飞机性能是基于标准条件下的空气密度,应用到空气密度高度的这个性能数据可能和高度计指示不一致。在高于或者低于标准的条件下,这些高度不能直接按高度计来计算。

密度高度先通过首次测得的压力高度来计算,然后根据非标准温度的变化来修正这个高度值。由于密度直接随压力而变化,随温度相反地变化,允许密度变化的时候一个给定的压力高度可能存在于很大范围的温度内。然而,一个已知的密度会在任何一个温度和压力高度下发生。当然,空气的密度对飞机和发动机性能有明显的影响。不管飞机运行的实际高度是多少,它都会表现出好像它运行在一个等于当前密度高度的高度上。例如,当设定为 29.92in Hg 时,高度计可能指示压力高度为 5000ft。根据飞机飞行手册/飞行员操作手册,飞机在标准温度条件下起飞时的地面滑跑可能要求距离为 790ft。

如果温度是标准之上的 20℃,空气的膨胀提高了密度高度。使用表格或者图表中的温度修正数据或者用计算机算出密度高度,可能发现密度高度是 7000ft,需要的地面滑跑距离可能要接近 1000ft。

空气密度受高度、温度和湿度变化的影响。高密度高度指的是稀薄空气,而低密度高度指的是稠密的空气。形成高密度高度的条件是高海拔高度、低大气压力、高温、高湿度或者这些因素中的某些因素的组合。低海拔高度、高大气压力、低温和低湿度是低密度高度的明显预兆。

使用飞行计算机,密度高度可以通过输入压力高度和飞行高度上的外部空气温度来计算。密度高度也可以参考图 3-3 和图 3-4 来计算。

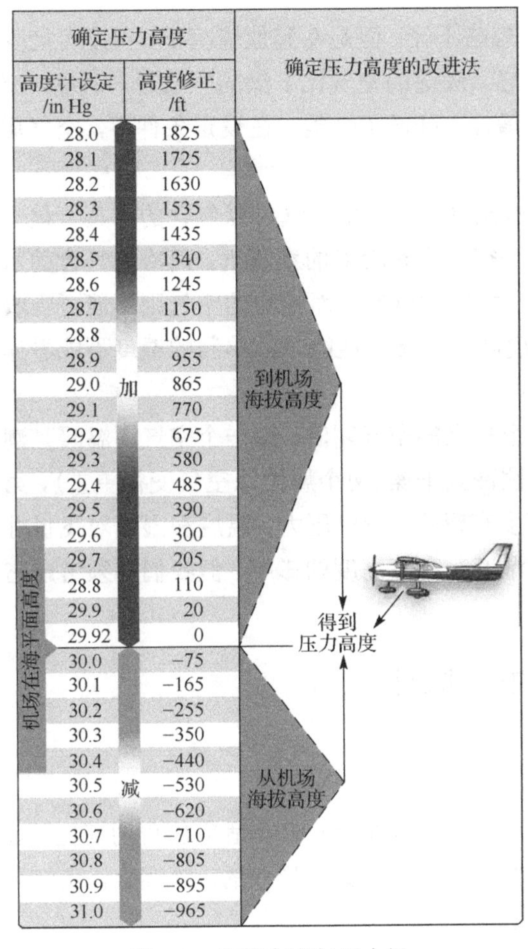

确定压力高度		确定压力高度的改进法
高度计设定/in Hg	高度修正/ft	
28.0	1825	
28.1	1725	
28.2	1630	
28.3	1535	
28.4	1435	
28.5	1340	
28.6	1245	
28.7	1150	
28.8	1050	
28.9	955	
29.0	865	
29.1	770	
29.2	675	
29.3	580	
29.4	485	
29.5	390	
29.6	300	
29.7	205	
28.8	110	
29.9	20	
29.92	0	
30.0	−75	
30.1	−165	
30.2	−255	
30.3	−350	
30.4	−440	
30.5	−530	
30.6	−620	
30.7	−710	
30.8	−805	
30.9	−895	
31.0	−965	

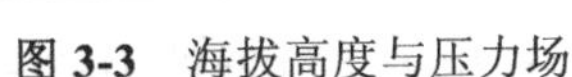

图 3-3　海拔高度与压力场

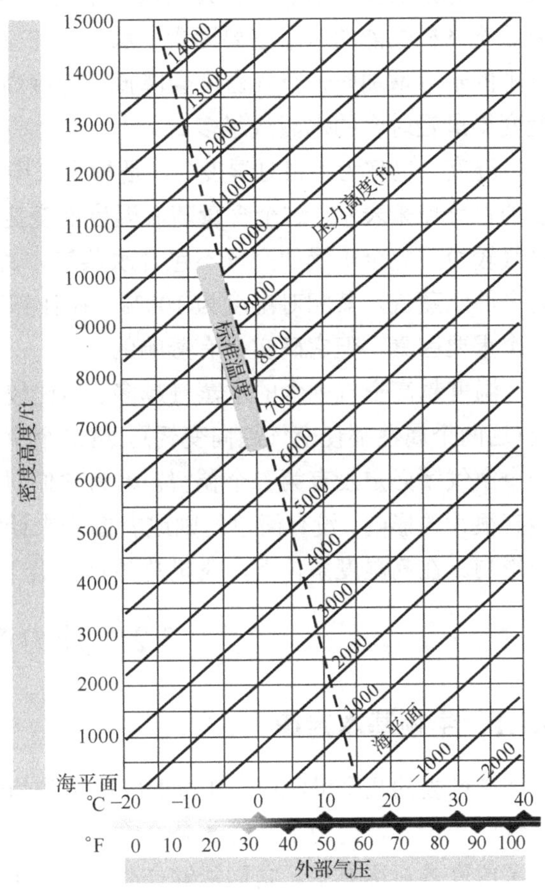

图 3-4　密度高度

3.1.4　压力对密度的影响

空气可以压缩或膨胀。当空气被压缩时，一定的体积就可以包含更多质量的空气。相反地，当作用于一定体积的空气的压力降低时，空气就会膨胀，占据更大的空间。也就是说，原来的空气柱在低压力时包含的空气质量更少，即密度降低了。实际上，密度和压力成比例。如果压力增加一倍，密度也增加一倍；如果压力降低，密度就会降低。这个结论只有在恒温时才成立。

3.1.5　温度对密度的影响

物质温度增加密度就会降低。相反，温度降低密度会增加。因此，空气密度和温度成反比。这个结论只有在恒压时成立。在大气中，温度和压力都随高度增加而降低，但对密度有相反的影响。然而，高度增加时压力的明显快速下降通常是主要的影响。因此，飞行员可以认为密度随高度增加而降低。

3.1.6　湿度(潮湿)对密度的影响

前面介绍的是基于空气为完全干燥的假设之上。实际上，空气从来不是完全干燥的。在大气中悬浮的少量水蒸气在某些情况下几乎可以忽略不计，但是在其他情况下，湿度可能变成影响飞机性能的一个重要因素。水蒸气比空气轻，潮湿的空气比干燥的空气轻。所以，空气中的水分增加时，空气密度会降低，密度高度增加，降低了性能。在设定条件下，空气质量变得最轻时含的水蒸气最多。

湿度，也称为"相对湿度"，是指大气中的水蒸气含量，用空气可以包含的水蒸气最大量的百分比来表示。这个含量随着温度而变化，暖空气含有较多的水蒸气，而冷空气含的水蒸气较少。完全干燥的空气不包含水蒸气，其相对湿度为 0%，而饱和的空气则不能再吸收更多的水蒸气，其相对湿度为 100%。在计算密度高度和飞机性能时，单独的湿度不能看作一个重要因素，但它确实是有影响的。

温度越高空气中含的水蒸气越多。当比较两个独立的空气团时，第一个空气团温暖且潮湿(这两个属性都使空气趋向变轻)，第二个空气团冷且干燥(两个属性让空气变得更重)，第一个空气团必定没有第二个稠密。因为它们对密度有影响，所以压力、温度和湿度对飞机性能有很大的影响。没有简单规则或者图表来计算湿度对密度高度的影响，但我们必须考虑它的影响。在高湿度条件下，飞机总体性能会下降。

3.2　升力产生的原理

3.2.1　牛顿基本定律

在过去的几个世纪中，升力形成理论作为物理界的基本理论一直在适应历史的变化。虽然这些理论看似适用于升力的各个方面，但是没有描述升力是如何形成的。事实上，我们必须考虑许多对称翼型，它们能够产生显著的升力。

在没有人成功进行第一次飞行器飞行之前，关于飞行中作用于飞行器的力的基本物理规律已经在假设的基础上建立起来了。这些物理定律的使用，催生了 17 世纪初在欧洲开始的科学革命。在宇宙以人们可预见方式运行理论的驱动下，许多哲学家、数学家、自然科学家、发明家用他们毕生的精力去揭开宇宙的秘密。其中最著名的是艾萨克·牛顿爵士，他不仅提出了万有引力定律，还描述了运动的三个基本定律。

牛顿第一定律：任何物体都保持静止状态或匀速直线运动，直到外力迫使它改变运动状态为止。

这意味着没有什么能让物体开始或停止运动，除非外力迫使它这样做。一架飞机停在停机坪上，会保持静止，除非施加一个足够大的力来克服惯性。一旦飞机开始运动，惯性会使它保持运动，克服施加在飞机上的其他力。这些力可推动其运动，可减慢其运动速度，也可改变其运动方向。物质加速度的大小跟其所受的合外力成正比，跟物体的质量成反比，加速度的方向跟合外力的方向相同。

牛顿第二定律：力等于动量的变化率。对于恒定质量，力等于质量乘以加速度。

当一个物体受到一个恒定力的作用时，其加速度与物体的质量成反比，并与所施加的力成正比。这里所涉及的就是克服牛顿第一定律的惯性的因素。它包含方向和速度的改变，且有两层含义：从静止到运动(正加速度)和从运动到停止(负加速度或者减速)。

牛顿第三定律：每一个作用力都会有一个反作用力，且大小相等，方向相反。

飞机上，螺旋桨转动向后推动空气，所以空气向相反的方向推螺旋桨，使飞机前进。在喷气式飞机上，引擎向后推动热空气气流，气流作用于引擎，通过等大小的作用力反向推动引擎，使得飞机前进。

3.2.2　马格努斯效应

1．空气经过不旋转的圆柱体

如果空气经过一个不旋转的圆柱体，经过圆柱体上面和下面的气流是一样的，力也是一样的，如图 3-5(a)所示。

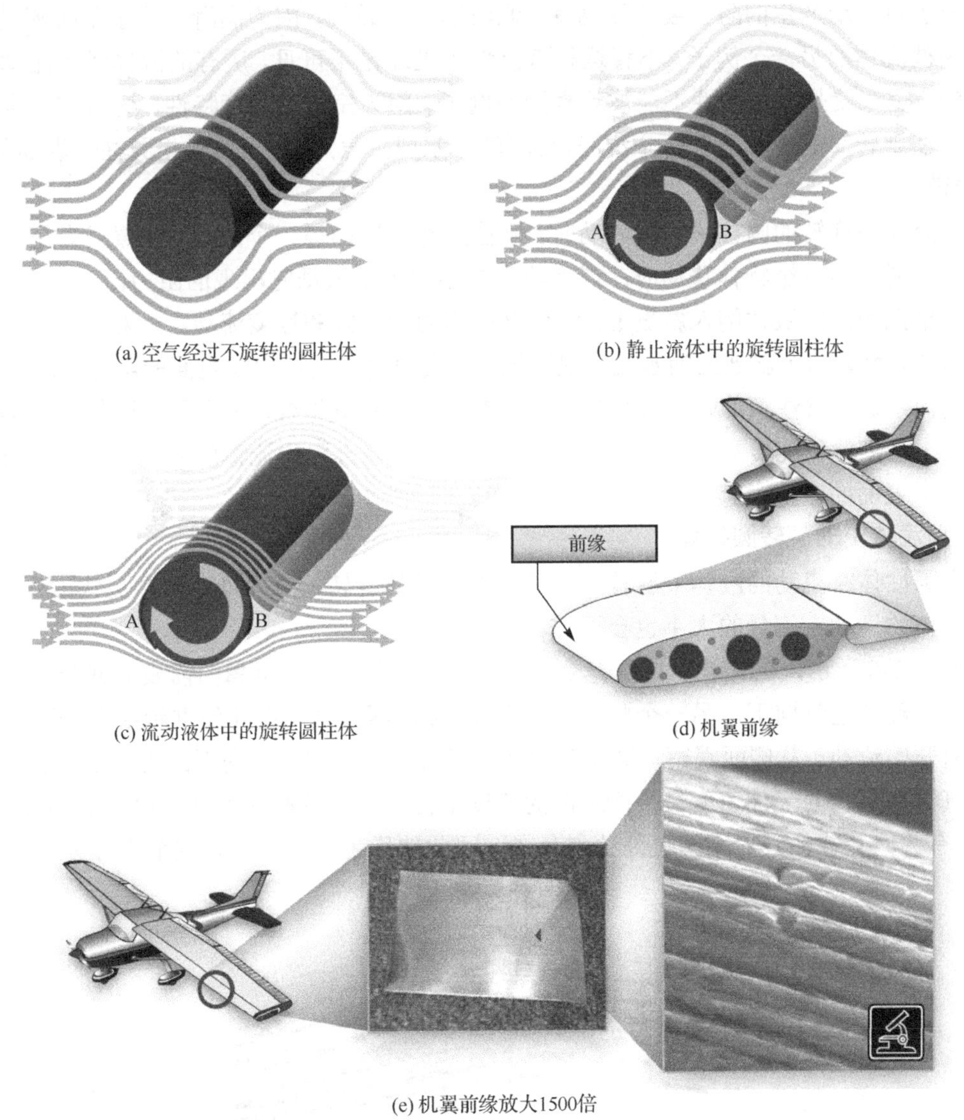

(a) 空气经过不旋转的圆柱体

(b) 静止流体中的旋转圆柱体

(c) 流动液体中的旋转圆柱体

(d) 机翼前缘

(e) 机翼前缘放大1500倍

图 3-5　(a)表明一个正常的循环，(b)描述了在旋转筒顶部增加的气流，(c)表明当旋转筒在气流中移动时气流速度显著增加，(d)是一种典型的航空级铝合金，在(e)所述的飞机结构如机翼和机翼前缘中使用

2．旋转的圆柱体在静止的流体中

在图 3-5(b)中，从侧面观察圆柱体顺时针旋转进入液体中时，圆柱体的旋转影响了自己周围的流体。黏度和摩擦两个因素引起的阻碍使旋转的圆柱体周围的流动与静止的圆柱体周围的流动不一样。

3．黏度

黏度是流体或半流体的一种性质，其能阻碍流动。在一定程度上，由于流体分子相互黏附，使流动阻力可衡量。高黏度流体阻碍流动；低黏度流体容易流动。

把等量的水和油倒入两个相同的斜坡上可以显示黏度的不同。水似乎自由流动，而油的流动速度要慢得多。

分子对运动产生的阻力要依靠黏度。润滑油非常黏稠，因为润滑油分子阻抗流动。热熔岩是黏稠流体的另一个例子。不管这种抵抗性可见还是不可见，所有的流体对于流动都是有黏性的。空气就是一个黏性不能被观察到的例子。因为空气的黏度特性，这将在一定程度上阻流。在旋转筒的情况下在一个沉浸流体(油、水或空气)中，流体阻止圆柱体表面的流动。

4．摩擦

当流体在旋转的圆柱体周围流动时，摩擦是起作用的另一个因素。摩擦阻力是一个面或者物体在遇到另一个物体移动时，或者流体在物理表面流动时出现的阻力。

如果把相同的液体倒入斜坡上，它们会以相同的方式和速度流动。如果一个斜坡的表面有小石子，那么两个斜坡上的水流明显不同。表面粗糙的斜坡由于表面的阻力(摩擦)阻碍了液体的流动。值得注意的是，所有的表面，不管他们光滑与否，都会阻止液体的流动。在微观层面上，机翼和旋转圆柱的表面都有一定的粗糙度，会阻碍流体的流动。空气流动速度的降低是由表面的摩擦引起的。

5．运动流体中的旋转圆柱体

当圆柱体在流动的液体中旋转时，液体在圆柱体旋转的方向形成一个更快的循环流动。增加流体的运动，流量的大小也会增加。

如图 3-6 所示，在 A 点，这里有个驻点，气流在此处撞上机翼的前表面并分开；部分空气向上，部分空气向下。另一个驻点存在于 B 点，两股气流在此处重新结合，以相同的速度重新开始流动。从侧面观察，机翼头部产生了升力而尾部产生向下的力。在图 3-6 的情况下，气流在机翼的上部速度最快而底部的速度最慢。由于这些速度都和研究的物体(这里为机翼)有关，并非在升力系统之外独立存在，因此被称为局部速度。这个概念可以较容易地

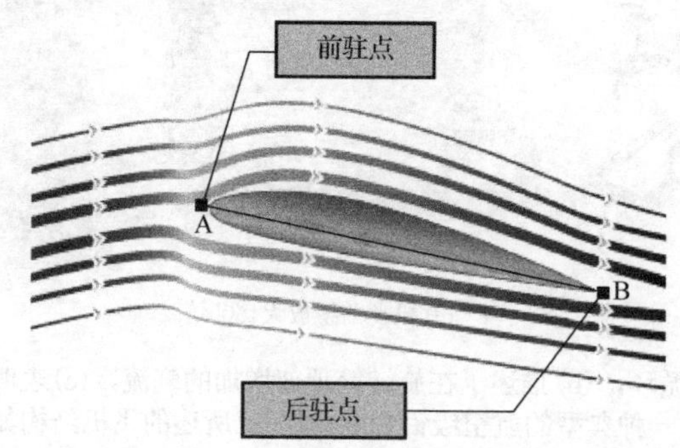

图 3-6　发生在机翼的空气循环，前驻点是低于机翼前缘，而后驻点超出了后缘

应用到机翼或其他升力面。因为在机翼的上下表面气流速度不同，这就导致了机翼下表面压力大，而上表面压力小。这个压力差产生一个向上的力的物理现象称为马格努斯效应，即一个物体的旋转会影响流过它的气体的路径。

3.2.3 伯努利效应

伯努利压力差原理的实践应用是文丘里管。文丘里管的入口比喉部(收缩点)直径大，出口部分的直径和入口一样大。在喉部，气流速度增加，压力降低；在出口处气流速度降低，压力增加，如图 3-7 所示。

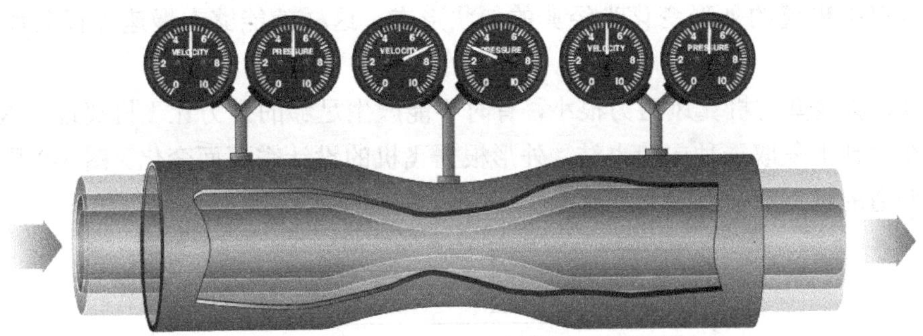

图 3-7 文丘里管中气压下降

3.3 机 翼 设 计

在讨论机翼的时候经常使用一条称为弦线的参考线，它是一条划过剖面图中两个端点前缘和后缘的直线。弦线到机翼上下表面的距离表示上下表面任意点的拱形程度。另一条参考线是从前缘划到后缘的线，叫中弧线，这条中弧线的所有点与上、下表面是等距的。典型机翼构造如图 3-8 所示。

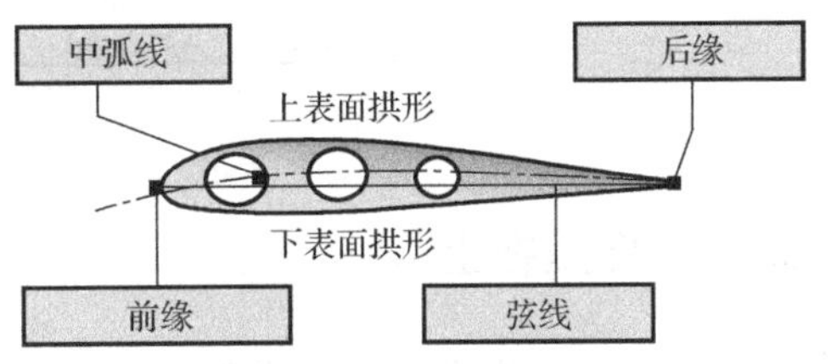

图 3-8 典型机翼构造

机翼的形状要依据空气对特定的物理定律的反应来设计。它从空气获得两种作用力：一种是从机翼下方空气产生的正压升力，另一种是从机翼上方产生的反向压力。

当机翼和其运动方向成一个小角度倾斜时，气流冲击相对较平的机翼下表面，空气被迫向下推动，所以导致了一个向上作用的升力，同时，冲击机翼前缘上曲面部分的气流向斜上方运动。也就是说，机翼产生一个作用于空气的力，迫使空气向下，也就提供了来自空气的相等的反作用力，迫使机翼向上。如果机翼的构造形状能够导致升力大于飞机的重量，飞机就可以飞起来了。

如果所有需要的力仅仅来自于机翼下表面导致的空气偏流，那么飞机就只需要一个类似风筝的平直机翼。然而，支撑飞机所需的升力平衡来自机翼上方的空气流动。这是飞行的关键。给机翼上表面产生的力和下表面产生的力指定一个具体的百分比既不精确，也没用处。

这些(来自上下表面的力以及它们的比例)都不是恒定值，它们的变化不仅取决于飞行条件，还和不同的机翼设计有关。

不同的机翼有不同的飞行特性。在风洞和实际飞行中测试了成千上万种机翼，但是没有发现一种机翼能够满足每一项飞行要求。每种飞机的重量、速度和用途决定了机翼的外形。很多年前人们就认识到产生最大升力的最有效率的机翼是一种有凹陷的下表面的勺状机翼。后来还认识到作为一种固定的设计，这种类型的机翼在产生升力的时候牺牲了太多的速度，因此不适合于高速飞行。工程技术的发展使今天的高速喷气机又开始利用勺状机翼的高升力特性这个优势。前缘(Kreuger)襟翼和后缘(Fowler)襟翼从基本机翼结构向外延伸时，直接把机翼的外形变化为经典的勺状形态，这样就能够在慢速飞行条件下产生很大的升力。

另外，流线型的机翼风阻力很小，有时不能产生足够的升力让飞机离地。这样，现代飞机机翼在设计上采取一种中庸办法，外形根据飞机的设计需要而变化。图 3-9 显示了一些设计上很普通的机翼剖面。

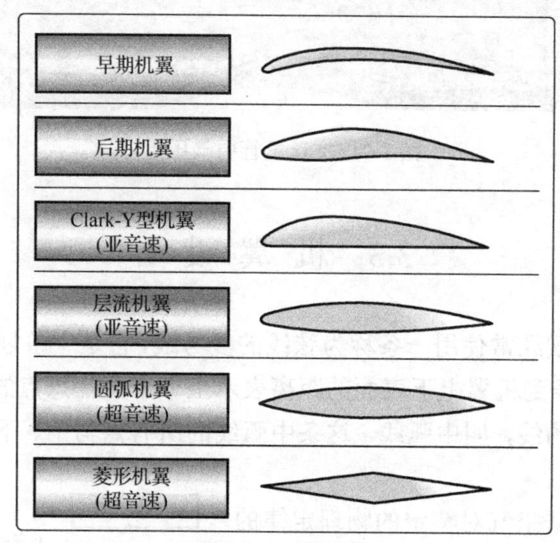

图 3-9　机翼设计

3.3.1　低压在上

在风洞试验或者实际飞行中，机翼仅仅是插入空气流中的流线型物体。如果机翼剖面呈泪滴状，流过机翼上下表面两边的空气速度和压力的变化是一样的。如果泪滴状机翼沿纵向切去一半，就可以产生构成基本机翼剖面的外形。如果机翼有倾角，气流就以一个角度(迎角)冲击它，由于上表面的弯曲引起运动距离的增加，导致机翼上表面移动的空气分子就被迫比沿下表面移动的分子更快。速度的增加减少了机翼上部的压力。

3.3.2　高压在下

机翼下方的压力条件决定升力的大小。由此，机翼下方的正压力在迎角较大时也相应增加。但是气流的另一方面也必须考虑。在靠近前缘的点，实际上气流是停滞的(停滞点)，然后逐渐地增加速度。在靠近尾缘的某些点，速度又变到和机翼上表面的速度相同。遵循伯努利压力差原理，机翼下方的气流速度较慢，产生了一个支撑机翼的正压力，当流体速度下降时，压力必定增加。由于机翼上下表面压力差的增加，因此机翼上的总升力增加。无论何时机翼产生的升力中伯努利压力差原理和牛顿定律都生效。

3.3.3 压力分布

从在风洞模型和全尺寸飞机上所做的试验来看，我们认为在不同迎角的机翼表面气流中，表面不同区域的压力有负的(比空气压力小)，也有正的(比空气压力大)。上表面的负压产生的力比下表面空气冲击机翼产生的正压得到的力更大。图 3-10 给出了三个不同迎角时沿机翼的压力分布。任何既定迎角的压力变化的平均值指中心压力(CP)。空气动力通过这个压力中心发生作用。较大迎角时压力中心前移，小迎角时压力中心后移。在机翼结构的设计中，压力中心的移动是非常重要的，因为它影响大迎角和小迎角时作用于机翼结构上的空气动力负荷的位置。飞机的航空动力学平衡和可控制性通过改变压力中心来调整。

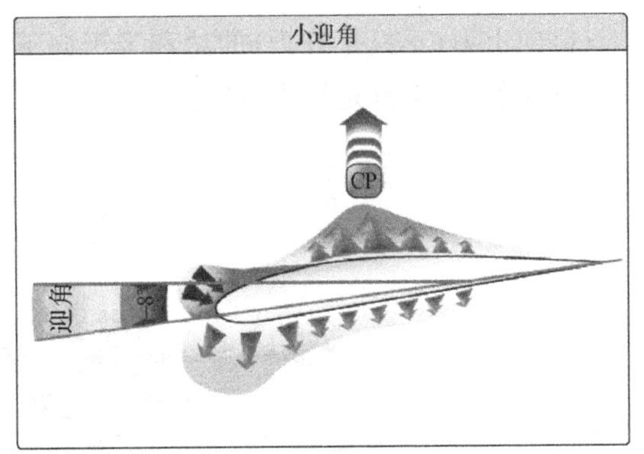

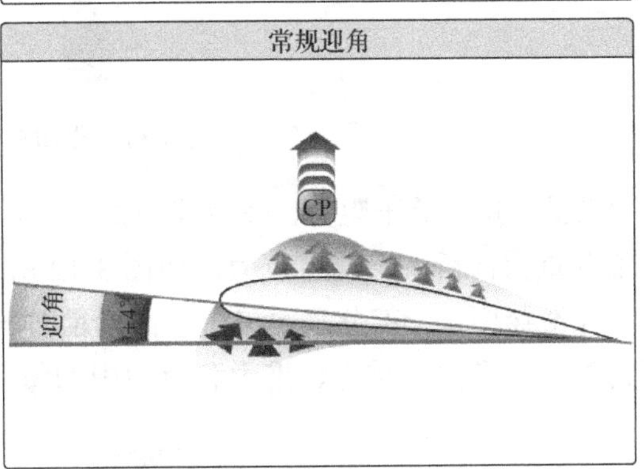

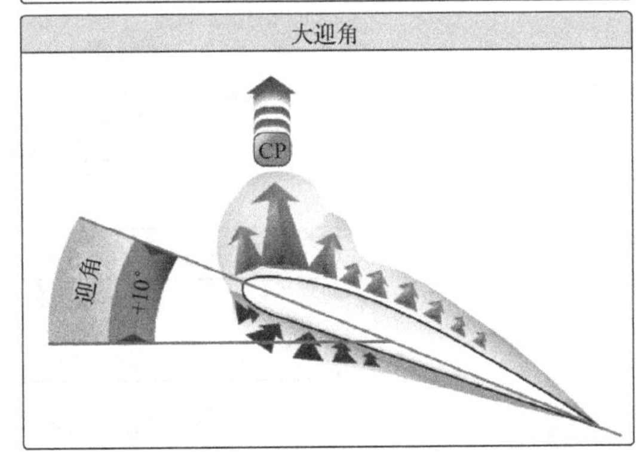

图 3-10 机翼上面的压力分布和中心压力随着迎角的变化

3.4 飞 行 原 理

在稳定的飞行中，这些相反作用的力的和总是为零。在稳定的直线飞行中没有不平衡的力(牛顿第二定律)。牛顿第三定律指出，相互作用的两个物体之间的作用力和反作用力总是大小相等，方向相反。无论飞机是水平飞行、爬升或者下降，这都是成立的。

但是也不等于说四个力总是相等的。这仅仅是说成对的反作用力大小相等，因此各自抵消对方的效果。如图 3-11 所示，图中的推力、阻力、升力和重力四个力矢量大小相等。通常解释说明推力等于阻力，升力等于重力。尽管这个表述基本正确，但还是容易让人误解。一定要明白在直线的、水平的、非加速飞行状态中，相反作用的升力和重力是相等的，但是它们也大于相反作用的推力和阻力。因此，在稳定的飞行中：

(1) 向上力(不只是升力)的总和等于向下力(不只是重力)的总和。

(2) 向前力(不只是推力)的总和等于向后力(不只是阻力)的总和。

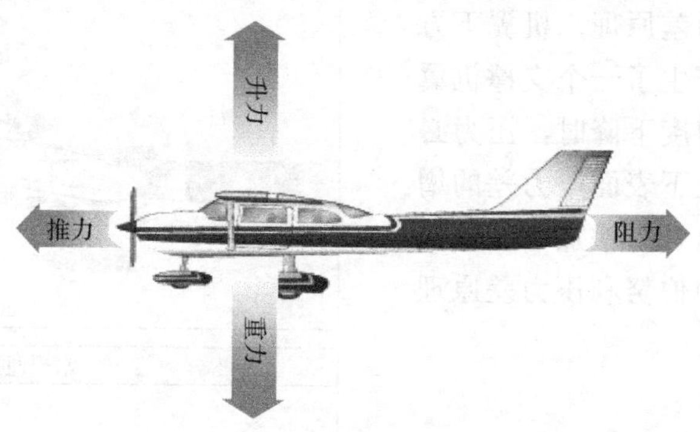

图 3-11　作用在飞机上的力的关系

公式“推力等于阻力，升力等于重力”可解释为：在爬升中部分推力向上，表现为升力，而部分重力向后，表现为阻力，如图 3-12 所示。

在滑翔中，一部分重力矢量方向向前，表现为推力。换句话说，在飞机航迹不是水平的任何时候，升力、重力、推力和阻力中的每一个都会分解为两个分力。

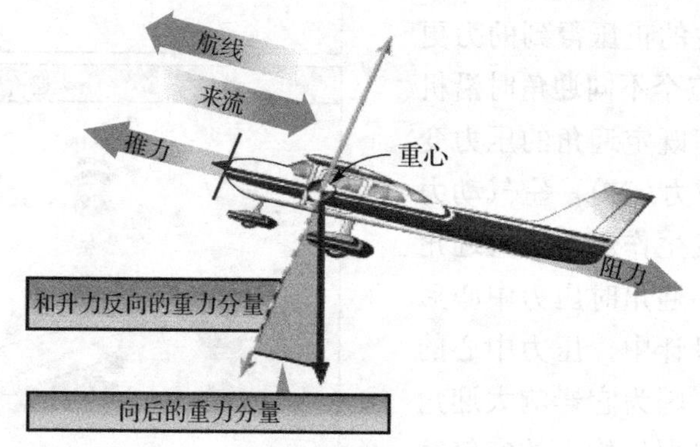

图 3-12　稳定爬升中力的矢量图

1．推力

飞机开始移动前，必须施加推力，且推力应大于阻力。飞机持续飞行，速度增加，直到推力和阻力相等。为了维持恒定的空速，就像升力和重力必须保持相等以维持稳定的飞行高度一样，推力和阻力必须保持相等。假设在平直飞行中，引擎功率降低，推力就会减少，飞机速度就会下降。只要推力小于阻力，飞机就会一直减速，直到它的空速不足以支持飞行。

同样，如果引擎的动力增加，推力比阻力大，空速就增加。只要推力一直比阻力大，飞机就一直加速。当阻力等于推力时，飞机以恒定的空速飞行。

2．阻力

阻力是抵抗飞机在空气中运动的力。阻力有两种基本类型：废阻力和诱导阻力。第一个称为寄生，因为它永远对飞行没有帮助；第二个是由机翼产生升力的结果所导致的。

3. 重力

重力是趋向把所有物体朝地球中心拽的拉力。重心可以看成是飞机的所有重量都集中于所在的一点。如果飞机的重心恰好得到支持，飞机就会平衡在任何位置。我们应该重视重心对飞机的重要性，因为它的位置对稳定性有极大的影响。

重心的位置通过每架飞机的总体设计来确定。设计者要确定压力中心会移动多大距离。相应地，他们会把重心固定在飞行速度下的压力中心前面，这是为了提供足够的恢复时间以保持飞行平衡。

4. 升力

飞行员可以控制升力。随着操纵杆的前后移动，迎角就会改变。当迎角增加时，升力增加(假设其他因素不变)。当飞机到达最大迎角时，升力开始快速降低。这就是失速迎角，或者叫紊流点。图 3-13 记录了升力系数 C_L 随着迎角的增大而增大，直至失速迎角后迎角变化很小，升力系数快速减小。

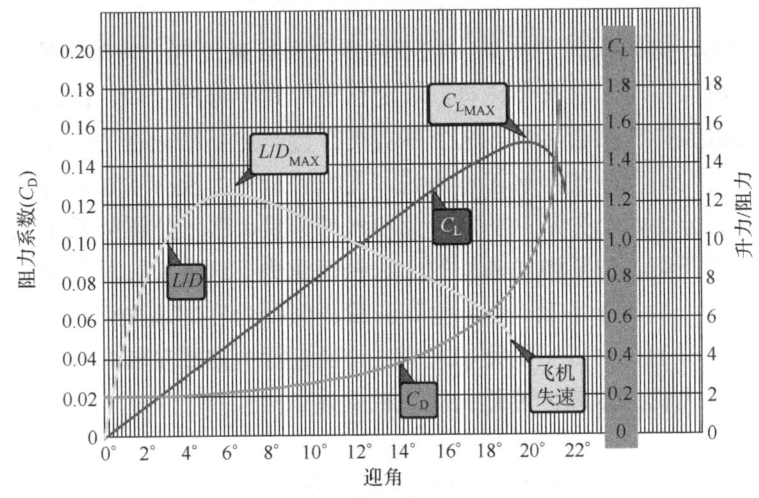

图 3-13　升力系数随迎角变化

5. 转弯受力

如果从前面看一架平直飞行的飞机，如图 3-14 所示，假设作用于飞机上的力可见，那么两个力(升力和重力)明显可见。如果飞机处于倾斜状态，可以明显地看到升力不再正好和重力方向相反，升力作用在倾斜的方向上。实际情况是，当飞机倾斜时，升力作用方向是朝转弯的中心且向上的，这是考虑飞机转弯时要记住的一个基本事实。

牛顿第一定律(即惯性定律)指出，任何物体都要保持静止或者匀速直线运动状态，直到有其他的力作用于这个物体。飞机和任何其他运动物体类似，需要有一个侧向力使它转弯。在一个正常的转弯中，这个力是通过飞机的倾斜得到的，这时升力是向上和向内作用的。转弯时的升力被分解为两个相互成直角的分量。竖直作用的分力和重力成对，称为垂直升力分量。另一个是水平的指向转弯的中心，称为水平升力分量，或者叫向心力。这个水平方向的力把飞机从直线航迹拉动到转弯航迹上。离心力和飞机转弯时的向心力方向相反，大小相等。这就解释了为什么在正常转弯时使飞机转弯的力不是方向舵施加的。方向舵是用来纠正直线航行时机头和机尾之间的航线误差。一个好的转弯就是机头和机尾沿着相同的轨迹。如果转弯的时候没有方向舵，机头就会偏出转弯轨迹。方向舵就是靠相对气流将机头带回飞行路线的。

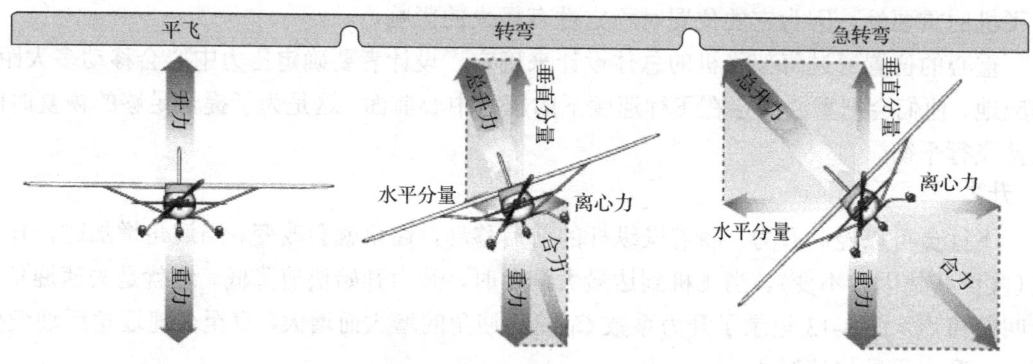

图 3-14　正常转弯中的受力

飞机为了转弯，必须倾斜。如果飞机不倾斜，那么就没有让它偏离原来直线航向的力。反过来说，当飞机倾斜时，它就会转弯，假定它不滑到转弯的一侧。良好的方向控制基于一个事实：只要飞机倾斜它就会转弯。这个事实飞行员一定要牢记在心，特别是要保持飞机平直飞行时。单就飞机的倾斜使得它转弯来说，飞机的总升力没有增加。然而就像提到过的一样，倾斜时的升力分为两个分量：垂直分量和水平分量。这一分解降低了抵消重力的力，进而飞机的高度就会下降，需要增加额外的力来抵消重力。这是通过增加迎角来实现的，直到升力的垂直分量再一次等于重量。由于垂直分量随倾斜角度的增加而降低，所以需要相应地增加迎角来产生足够的升力以平衡飞机的重力。当进行恒定高度转弯时，一定要记住升力的垂直分量必须等于飞机的重量才能维持飞机的高度。

3.5　飞机轴向和稳定性

轴向是通过飞机重心而想象出来的三条线。飞机的轴向可以看成飞机可以绕着它转动的假想轴。在三个轴的相交点，每一个轴都和其他两个轴成 90° 角。从飞机头部到尾部沿机身长度方向扩展的轴称为纵轴。从机翼到机翼的延伸轴称为横轴。垂直通过重心的轴叫垂直轴。飞机在飞行中，无论什么时候改变飞行高度和位置都绕三个轴向中的一个或者多个旋转。如图 3-15 所示。

飞机关于纵轴的运动和船舶从一边到另一边的侧滚运动相似。实际上，用来描述飞机关于三个轴运动的名字起源于航海术语。它们被用作航空术语是因为飞机和航海运动的相似性。飞机纵轴固定后的运动被称为“侧滚”，横轴固定时的运动叫“俯仰”，飞机垂直轴固定后的运动叫“偏航”。偏航是飞机头水平(左右)运动。

稳定性是飞机纠正那些可能改变它的平衡条件的内在品质，以及返回或继续在原航迹上飞行的能力。这是飞机的一个主要设计特性。飞机飞行的航迹和高度仅受飞机的空气动力学特性、推进系统和结构强度限制。这些限制决定了飞机的最大性能和机动性。如果飞机要提供最大效用，在这些限制的全部范围内它必须是安全可控的，且不超出飞行员的强度和要求额外的飞行能力。如果飞机沿任意航迹笔直、稳定地飞行，那么作用于飞机的力必定是静态平衡的。任何物体的平衡受到破坏后的反应和稳定性有关。稳定性有两种：静态稳定性和动态稳定性。

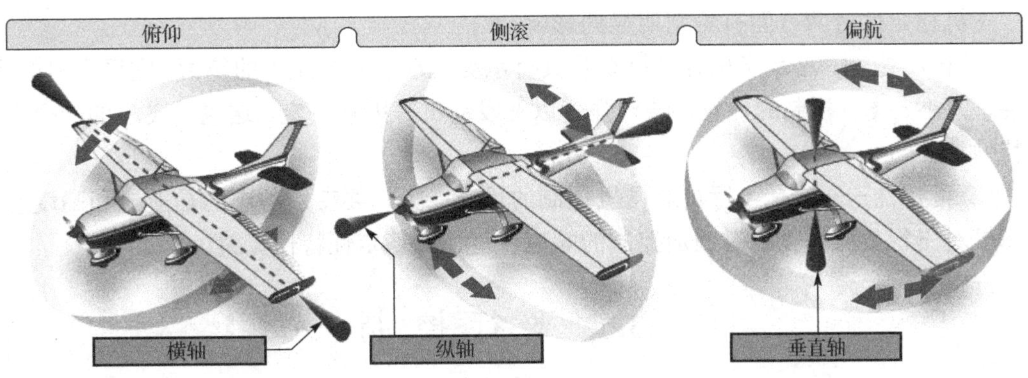

图 3-15 飞机的轴

1. 静态稳定性

静态稳定性是指飞机平衡条件破坏后的初始趋势或运动方向。在航空中，它是指飞机在迎角改变、侧滑或者倾斜时候的初始反应，如图 3-16 所示。

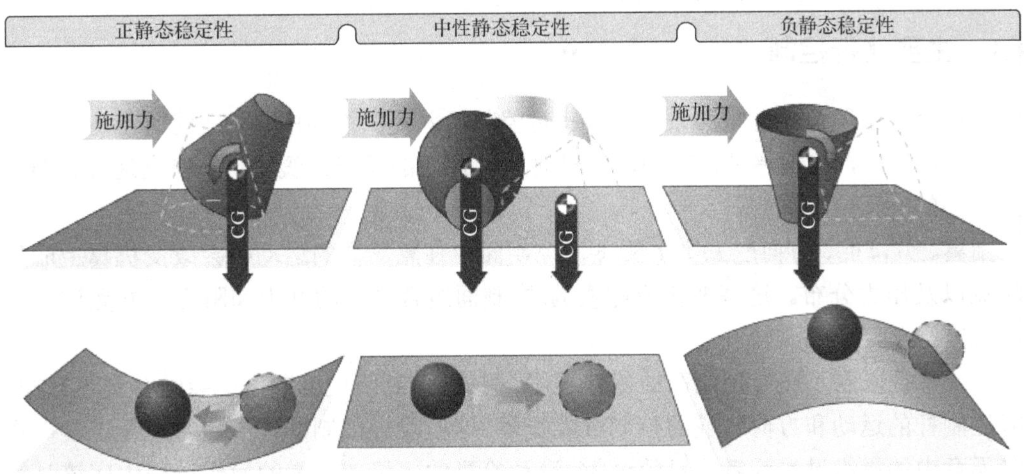

图 3-16 静态稳定性类型

(1) 正静态稳定性——飞机平衡被破坏后返回到原来平衡状态的最初趋势。

(2) 中性静态稳定性——飞机平衡被破坏后维持在一个新条件的最初趋势。

(3) 负静态稳定性——飞机平衡被破坏后持续偏离原来平衡状态的最初趋势。

2. 动态稳定性

静态稳定性是指飞机展现出来的原先状态受扰动后回复平衡的初始趋势。有时候，初始趋势和总体趋势不同或者相反，因此两者必须区别。动态稳定性是飞机从给定迎角、侧滑或者倾斜状态受到扰动后随时间的反应。

(1) 正动态稳定性——随着时间变化，偏移物体的振幅减小，且因为是正的，偏移物体回复到平衡状态。

(2) 中性动态稳定性——一旦发生偏移，物体的振幅既不减小也不增大。一个磨损的汽车减震器体现出这种趋势。

(3) 负动态稳定性——随着时间变化，偏移物体的运动加剧而且变得更加离散。

飞机稳定性对两方面有特别明显的影响：

(1) 机动性——这是飞机容易机动且具有承受机动引发的压力的能力。它受飞机的重量、惯性、大小、飞行控制的位置、结构强度以及发动机等因素决定。这也是飞机的一个主要设计特性。

(2) 可控性——这是飞行员对飞机控制的反应能力，要特别考虑的是航迹和高度态。它是飞机对飞行员操作飞机时施加控制的响应特性，和稳定性特性无关。

3.6 飞行控制

飞行器飞行控制系统分为主要飞行控制系统和辅助飞行控制系统。主要飞行控制系统包含飞行中要求的安全控制飞机的装置，如副翼、升降舵(或者安定面)以及方向舵。辅助飞行控制系统主要有襟翼、前缘装置、扰流板和配平系统。辅助飞行控制系统提升了飞机的性能特性，或者减轻了飞行员的过多控制力。

3.6.1 主要飞行控制

飞机控制系统都经过精心设计，以针对控制输入提供恰当的响应，同时营造一个很自然的感觉。低速时，控制感是偏软的，飞机对施加控制的反应是缓慢的。在高速飞行时，控制感是偏硬的，反应也更快。

副翼、升降舵、方向舵三个主要飞行控制面中任意一个的运动都会改变机翼上面和周围的气流以及压力分布。这些变化影响机翼/控制面组合产生的升力和阻力，并允许飞行员控制飞机围绕某三个轴旋转。

设计特征限制了飞行控制面的偏转程度。例如，控制停止机制可能会结合到飞行控制中，或者控制杆的运动和方向脚舵的控制可能受限。这些设计限制的目的是防止在正常操作时飞行员无意中的操纵过量或者飞机的过载。设计恰当的飞机在正常的操作过程中应该是稳定且容易控制的。控制面的输入导致三个轴向的旋转运动。飞机表现出来的稳定性类型也和三个轴向的旋转有关，如图 3-17 所示。

1. 副翼

副翼控制纵轴方向的侧滚。副翼安装在每一个机翼的后缘外侧，且运动方向彼此相反。副翼通过线缆、双臂曲柄和滑轮或推挽式管互相连接，然后相连到控制轮。

向右移动控制轮导致右侧副翼向上偏转，左侧副翼向下偏转。右侧副翼的向上偏转降低了机翼的拱形，使右侧机翼的升力降低。相应的左侧副翼的向下偏转增加了拱形幅度，使左侧机翼的升力增加。因此，左侧机翼的升力增加和右侧机翼的升力降低使飞机向右侧滚。

2. 逆偏转

向下偏转的副翼产生更大的升力，也产生更大的阻力，这个增加的阻力使飞机慢慢减速，使飞机头朝机翼上升的一侧偏转。从飞行员角度来说，偏转和倾斜的方向是相反的。逆偏转是由于左右机翼的阻力和速度上的差别造成的，如图 3-18 所示。

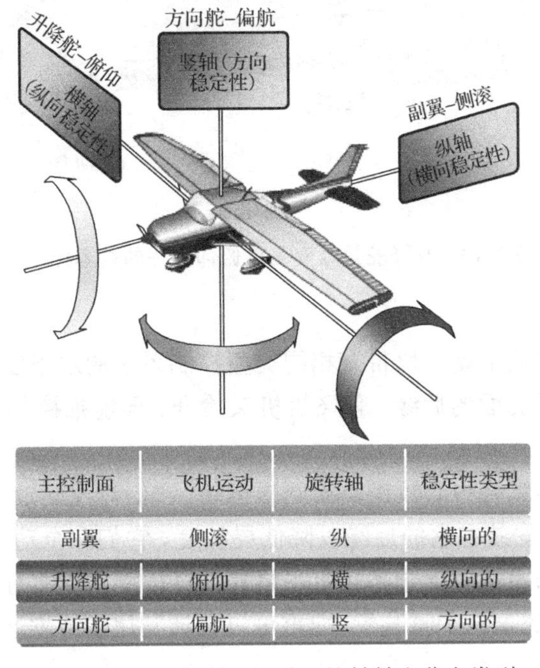

主控制面	飞机运动	旋转轴	稳定性类型
副翼	侧滚	纵	横向的
升降舵	俯仰	横	纵向的
方向舵	偏航	竖	方向的

图 3-17　飞行控制、运动、旋转轴和稳定类型

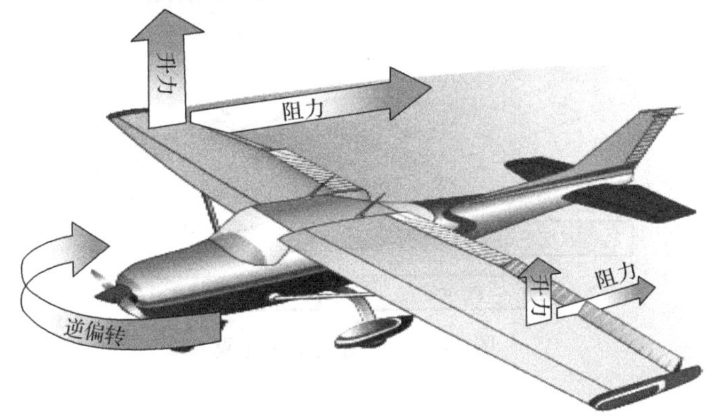

图 3-18　逆向偏转是由产生更多升力的外侧机翼上的大阻力造成的

3．升降舵

升降舵控制横轴的俯仰运动。类似小飞机上的副翼，升降舵通过一系列机械连杆机构连接到座舱中的控制杆。控制杆后移使升降舵面的后缘向上偏转。如图 3-19 所示的升降舵。

上偏升降舵位置减弱了升降舵的拱形，产生了一个向下的空气动力，它比平直飞行时的正常尾部向下的力要大。总体效果是导致飞机的尾部向下移动，机头上仰。俯仰运动围绕重心发生。俯仰运动的强度由重心和水平尾翼面的距离和水平尾翼上有效气动力决定。向前移动控制杆有相反的效果。这种情况下，升降舵的拱形度增加，水平安定面上产生的升力更多(尾部向下的力更小)。这就使得尾部向上移动，机头下俯。俯仰运动还是围绕飞机重心发生的。

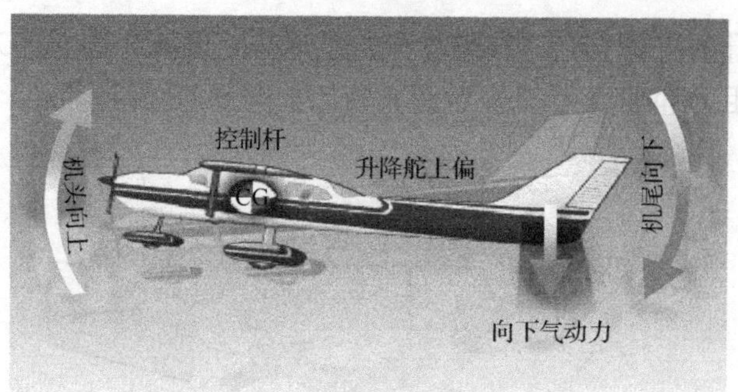

图 3-19　升降舵是改变飞机俯仰姿态的基本控制手段

4．全动式水平尾翼

全动式水平尾翼本质上是一块带有相同类型控制系统的水平安定面。当控制杆后拉时，它抬升了全动式水平尾翼面的后缘，使飞机机头抬升。向前推控制杆，使水平尾翼的后缘放低，机头向下俯。

因为全动式水平尾翼绕中心铰链点做回转运动，它们对控制输入和空气动力负载相当敏感。反作用伺服调整片安装在它的后缘以降低灵敏度。它们和全动式水平尾翼向同样的方向倾斜。因此需要更大的力来驱动全动式水平尾翼，这样不易受飞行员带来的过度控制影响。另外，在主翼梁的前面还配有配重装置。配重可以设计到尾部或安装在全动式尾翼片的前部，如图 3-20 所示。

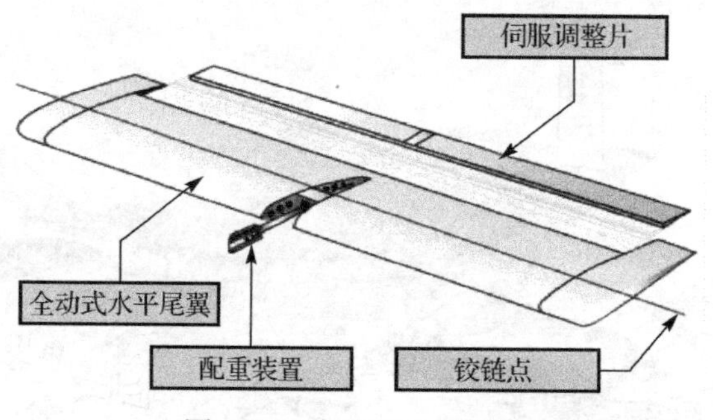

图 3-20　全动式水平尾翼

5．鸭式机翼

鸭式机翼设计利用了两种升力面的概念。鸭式机翼安装在主机翼前面，作为飞机的水平安定面。实际上，它是一种类似于常规后尾设计水平控制面的翼型。区别是鸭式机翼实际上产生升力，保持机头抬升，和后尾设计相反，后尾设计会在尾部施加向下的力来防止机头向下偏。

鸭式机翼设计可以追溯到早期的航空业，最有名的应用是在莱特兄弟飞行器。近些年，鸭式结构受到欢迎，并开始出现在较新的飞机上。鸭式布局包括两种类型：一种是水平控制面和正常的后尾设计有大约相同的尺寸，另外一种是大小差不多的控制面，翼型是串联翼配置的后安装式机翼。理论上认为鸭式机翼效率更高，因为在一定大小的升力的情况，利用水平控制面来帮助抬升飞机的重量，阻力更少。

6. 方向舵

方向舵控制飞机沿垂直轴的运动，这个运动称为偏航。和其他主要控制面类似，方向舵也是一个铰链到固定面的可运动面，在这种情况下它铰链到垂直安定面上。左右方向舵踏板的运动控制方向舵。当方向舵偏转到气流中时，会在相反的方向上施加水平方向的力，如图 3-21 所示。

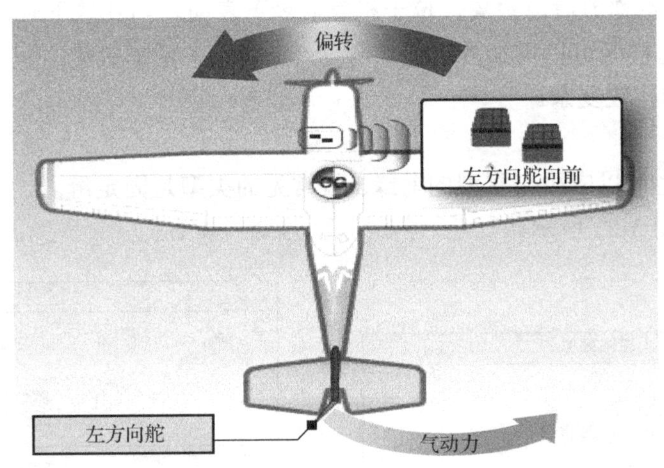

图 3-21 左方向舵压力的影响

通过踩踏左踏板，方向舵向左移动。这就改变了垂直安定面/方向舵周围的气流，产生一个侧向力，把尾部向右移动，使得飞机头向左偏航。方向舵有效性随速度而增加，因此在低速飞行时的大角度偏转和高速飞行时的小角度偏转能够提供需要的反作用力。对于螺旋桨驱动的飞机，流过方向舵的任何滑流都会增加它的有效性。

3.6.2 辅助飞行控制

1. 襟翼

襟翼是几乎所有飞机都使用的最常见的高升力装置。对任何设定的迎角，这些安装在机翼后缘的控制面既增加了升力又增加了诱导阻力。襟翼容许在高巡航速度和低着陆速度之间折中，因为它可以在需要的时候伸出，不需要的时候收回到机翼里。通常有四种襟翼，包括简单襟翼、分裂襟翼、开缝襟翼和福勒襟翼，如图 3-22 所示。

(1) 简单襟翼是四种襟翼中最简单的一种。它增加翼面弯度，导致一定迎角时的升力系数明显增加。同时它也大大地增加了阻力，而且把机翼压力中心向后移动，导致了机头的下俯运动。

(2) 分裂襟翼从机翼的下表面分离出来，它比简单襟翼产生的升力稍有增加。但是，由于在机翼后产生了紊乱的气流模式，所以产生的阻力更多。当襟翼完全伸出时，简单襟翼和分裂襟翼都产生高阻力，而升力增加不多。

(3) 开缝襟翼是现今飞机上最流行的襟翼。这种设计既用于小型飞机也用于大型飞机。开缝襟翼比简单襟翼和分裂襟翼明显提高了升力系数。对于小型飞机，铰链位于襟翼下表面的下面，当襟翼放下时，它在机翼的襟翼槽和襟翼前缘之间形成一个导气槽。当开缝襟翼放下时，来自下表面的高能量空气被输送到襟翼的上表面。来自导气槽的高能量空气加速了上表面边界层流，延迟了气流分离，提供了更高的升力系数。因此，开缝襟翼产生的最大升力系数比简单襟翼和分裂襟翼要高得多。开缝襟翼有很多类型，大飞机通常有双开缝襟翼，甚至是三开缝襟翼。这些襟翼使阻力最大限度地增加而不会出现襟翼上的气流分离和升力的减小。

(4)福勒襟翼也称为后退襟翼，位于机翼后缘下表面。当其放下时，可以一边向下偏转一边向后移动，这样既可以增加翼剖面的弯度，也能大大增加机翼面积。其升力系数可提高85%～95%，但结构较复杂。

2．前缘装置

高升力装置也可以应用到翼型的前缘。最常见的类型是固定缝翼、可动缝翼、前缘襟翼和前缘缝翼。如图 3-23 所示。

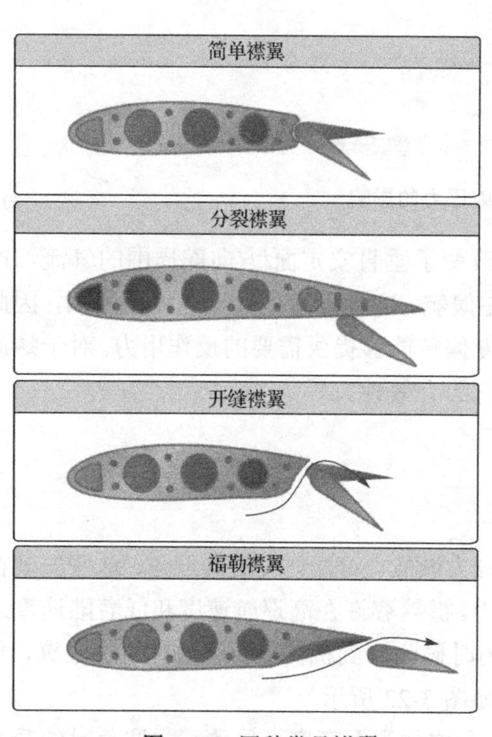

图 3-22　四种常见襟翼

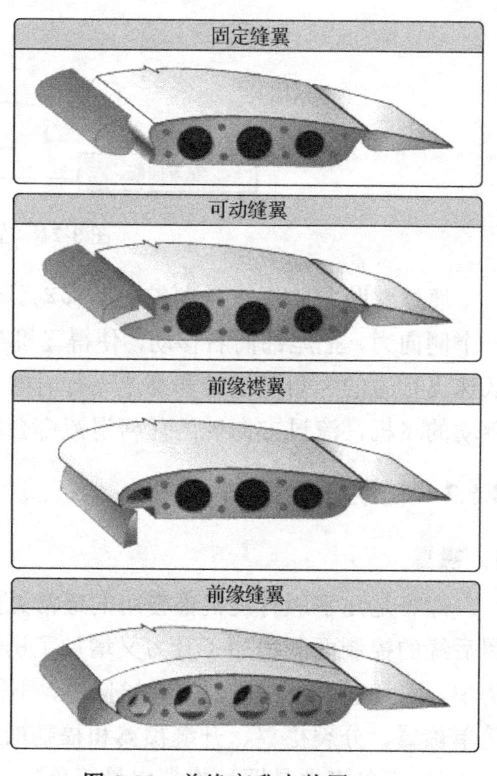

图 3-23　前缘高升力装置

固定缝翼把气流引导到机翼的上表面，延迟了大迎角时的气流分离。缝翼不增加机翼的弯度，但是让机翼获得更高的最大升力系数，因为在机翼到达一个更大的迎角之前失速被延迟了。

3．扰流板

在许多滑翔机和飞机上，被安装在机翼上的高阻力装置称为扰流板，用来扰乱平滑的气流、降低升力和增加阻力。有些飞机上扰流板用于侧滚控制，它的一个好处是消除了逆偏转。例如，要右转弯，右侧机翼上的扰流板抬起，损失一些升力，在右边产生更多的阻力。这时右边的机翼下降，飞机向右倾斜和偏航。两侧机翼同时使用扰流板会使飞机下降而速度不增加。扰流板用于帮助缩短着陆后的地面滑跑距离。通过损失升力，它们把重量转移到轮子上，提高了减速效果。下降过程中扰流板增加阻力并减少升力，如图 3-24 所示。

图 3-24 下降过程中扰流板增加阻力并减少升力

Chapter 4 Aircraft Systems

4.1 Powerplant

An aircraft engine, or powerplant, produces thrust to propel an aircraft. Reciprocating engines and turboprop engines work in combination with a propeller to produce thrust. All of these powerplants also drive the various systems that support the operation of an aircraft.

4.1.1 Reciprocating Engine

Most small aircraft are designed with reciprocating engines. The name is derived from the back-and-forth, or reciprocating, movement of the pistons which produces the mechanical energy necessary to accomplish work.

The engine designs can be further classified as:

(1) Cylinder arrangement with respect to the crankshaft—radial, in-line, V-type, or opposed.

(2) Operating cycle—two or four.

(3) Method of cooling—liquid or air.

Radial engines were widely used during World War II and many are still in service today. With these engines, a row or rows of cylinders are arranged in a circular pattern around the crankcase. The main advantage of a radial engine is the favorable power-to-weight ratio.

In-line engines have a comparatively small frontal area, but their power-to-weight ratios are relatively low. In addition, the rearmost cylinders of an air-cooled, in-line engine receive very little cooling air, so these engines are normally limited to four or six cylinders. V-type engines provide more horsepower than in-line engines and still retain a small frontal area. Continued improvements in engine design led to the development of the horizontally-opposed engine which remains the most popular reciprocating engines used on smaller aircraft. These engines always have an even number of cylinders, since a cylinder on one side of the crankcase "opposes" a cylinder on the other side.The majority of these engines are air cooled and usually are mounted in a horizontal position when installed on fixed-wing airplanes. Opposed-type engines have high power-to-weight ratios because they have a comparatively small, lightweight crankcase. In addition, the compact cylinder arrangement reduces the engine's frontal area and allows a streamlined installation that minimizes aerodynamic drag.

The main parts of a spark ignition reciprocating engine include the cylinders, crankcase, and accessory housing. The intake/exhaust valves, spark plugs, and pistons are located in the cylinders. The crankshaft and connecting rods are located in the crankcase. The magnetos are normally located on the engine accessory housing (Figure 4-1).

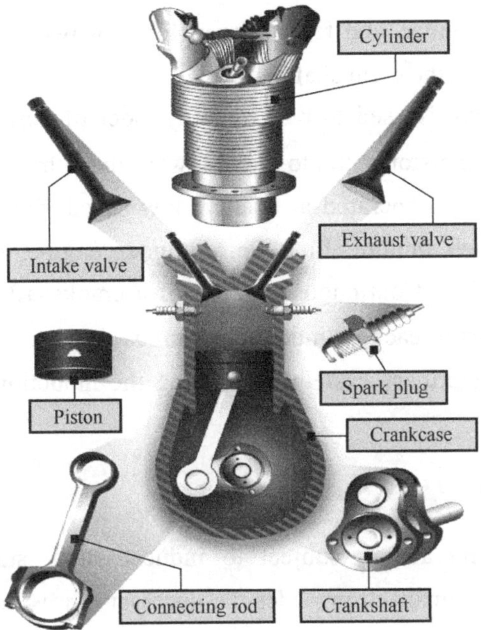

Figure 4-1 Main components of a spark ignition reciprocating engine

In a four-stroke engine the conversion of chemical energy into mechanical energy occurs over a four stroke operating cycle. The intake, compression, power, and exhaust processes occur in four separate strokes of the piston (Figure 4-2).

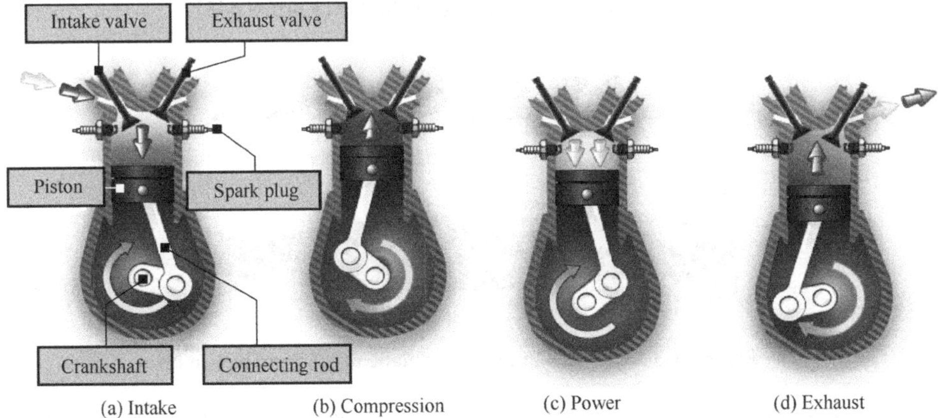

Figure 4-2 The arrows in this illustration indicate the direction of motion of the crankshaft and piston during the four-stroke cycle

(1) The intake stroke begins as the piston starts its downward travel. When this happens, the intake valve opens and the fuel/air mixture is drawn into the cylinder.

(2) The compression stroke begins when the intake valve closes and the piston starts moving back to the top of the cylinder. This phase of the cycle is used to obtain a much greater power output from the fuel/air mixture once it is ignited.

(3) The power stroke begins when the fuel/air mixture is ignited. This causes a tremendous

pressure increase in the cylinder, and forces the piston downward away from the cylinder head, creating the power that turns the crankshaft.

(4) The exhaust stroke is used to purge the cylinder of burned gases. It begins when the exhaust valve opens and the piston starts to move toward the cylinder head once again.

Even when the engine is operated at a fairly low speed, the four-stroke cycle takes place several hundred times each minute. In a four-cylinder engine, each cylinder operates on a different stroke (Figure 4-2). Continuous rotation of a crankshaft is maintained by the precise timing of the power strokes in each cylinder. Continuous operation of the engine depends on the simultaneous function of auxiliary systems, including the induction, ignition, fuel, oil, cooling, and exhaust systems.

4.1.2 Propeller

The propeller is a rotating airfoil, subject to induced drag, stalls, and other aerodynamic principles that apply to any airfoil. It provides the necessary thrust to pull, or in some cases push, the aircraft through the air. The engine power is used to rotate the propeller, which in turn generates thrust very similar to the manner in which a wing produces lift. The amount of thrust produced depends on the shape of the airfoil, the angle of attack of the propeller blade, and the revolutions per minute (rpm) of the engine. The propeller itself is twisted so the blade angle changes from hub to tip. The greatest angle of incidence, or the highest pitch, is at the hub while the smallest angle of incidence or smallest pitch is at the tip (Figure 4-3).

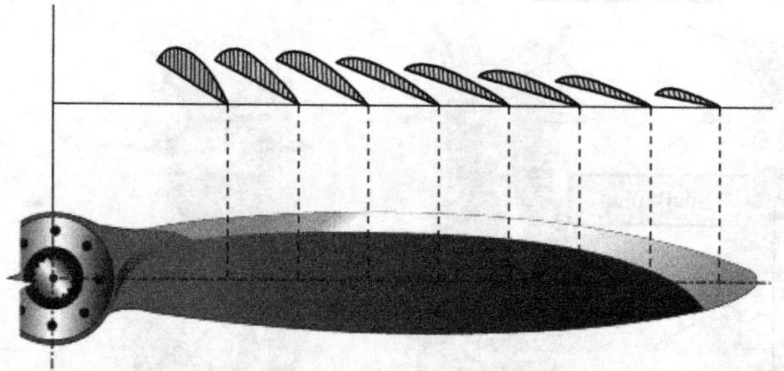

Figure 4-3 Changes in propeller blade angle from hub to tip

The reason for the twist is to produce uniform lift from the hub to the tip. As the blade rotates, there is a difference in the actual speed of the various portions of the blade. The tip of the blade travels faster than the part near the hub, because the tip travels a greater distance than the hub in the same length of time. Changing the angle of incidence (pitch) from the hub to the tip to correspond with the speed produces uniform lift throughout the length of the blade. A propeller blade designed with the same angle of incidence throughout its entire length would be inefficient because as airspeed increases in flight, the portion near the hub would have a negative angle of attack while the blade tip would be stalled (Figure 4-4).

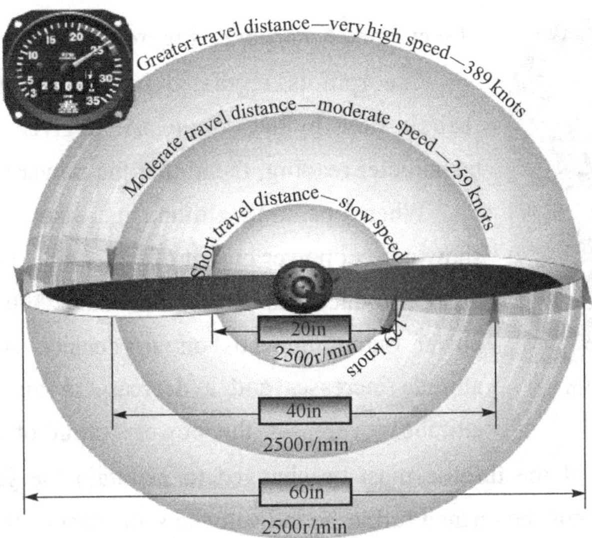

Figure 4-4 Relationship of travel distance and speed of various portions of propeller blade

Small aircraft are equipped with either one of two types of propellers. One is the fixed pitch, and the other is the adjustable pitch.

1. Fixed-Pitch Propeller

A propeller with fixed blade angles is a fixed-pitch propeller. The pitch of this propeller is set by the manufacturer and cannot be changed. Since a fixed-pitch propeller achieves the best efficiency only at a given combination of airspeed and rpm, the pitch setting is ideal for neither cruise nor climb. Thus, the aircraft suffers a bit in each performance category. The fixed-pitch propeller is used when low weight, simplicity, and low cost are needed.

There are two types of fixed-pitch propellers: climb and cruise. Whether the airplane has a climb or cruise propeller installed depends upon its intended use. The climb propeller has a lower pitch, therefore less drag. Less drag results in higher rpm and more horsepower capability, which increases performance during takeoffs and climbs, but decreases performance during cruising flight. The cruise propeller has a higher pitch, therefore more drag. More drag results in lower rpm and less horsepower capability, which decreases performance during takeoffs and climbs, but increases efficiency during cruising flight.

The propeller is usually mounted on a shaft, which may be an extension of the engine crankshaft. In this case, the rpm of the propeller would be the same as the crankshaft rpm. On some engines, the propeller is mounted on a shaft geared to the engine crankshaft. In this type, the rpm of the propeller is different than that of the engine.

In a fixed-pitch propeller, the tachometer is the indicator of engine power. A tachometer is calibrated in hundreds of rpm and gives a direct indication of the engine and propeller rpm. The instrument is color coded, with a green arc denoting the maximum continuous operating rpm. Some tachometers have additional markings to reflect engine and/or propeller limitations. The manufacturer's recommendations should be used as a reference to clarify any misunderstanding

Figure 4-5 Engine rpm is indicated on the tachometer

of tachometer markings (Figure 4-5).

The rpm is regulated by the throttle, which controls the fuel/air flow to the engine. At a given altitude, the higher the tachometer reading, the higher the power output of the engine.

When operating altitude increases, the tachometer may not show correct power output of the engine. For example, 2300 rpm at 5000ft produces less horsepower than 2300 rpm at sea level because power output depends on air density. Air density decreases as altitude increases and a decrease in air density (higher density altitude) decreases the power output of the engine. As altitude changes, the position of the throttle must be changed to maintain the same rpm. As altitude is increased, the throttle must be opened further to indicate the same rpm as at a lower altitude.

2. Adjustable-Pitch Propeller

The adjustable-pitch propeller was the forerunner of the constant-speed propeller. It is a propeller with blades whose pitch can be adjusted on the ground with the engine not running, but which cannot be adjusted in flight. It is also referred to as a ground adjustable propeller. By the 1930s, pioneer aviation inventors were laying the ground work for automatic pitch-change mechanisms, which is why the term sometimes refers to modern constant-speed propellers that are adjustable in flight.

The first adjustable-pitch propeller systems provided only two pitch settings: low and high. Today, most adjustable-pitch propeller systems are capable of a range of pitch settings.

A constant-speed propeller is a controllable-pitch propeller whose pitch is automatically varied in flight by a governor maintaining constant rpm despite varying air loads. It is the most common type of adjustable-pitch propeller. The main advantage of a constant-speed propeller is that it converts a high percentage of brake horsepower (BHP) into thrust horsepower (THP) over a wide range of rpm and airspeed combinations. A constant-speed propeller is more efficient than other propellers because it allows selection of the most efficient engine rpm for the given conditions.

An aircraft with a constant-speed propeller has two controls: the throttle and the propeller control. The throttle controls power output and the propeller controls and regulates engine rpm. This in turn regulates propeller rpm which is registered on the tachometer.

Once a specific rpm is selected, a governor automatically adjusts the propeller blade angle as necessary to maintain the selected rpm. For example, after setting the desired rpm during cruising flight, an increase in airspeed or decrease in propeller load will cause the propeller blade angle to increase as necessary to maintain the selected rpm. A reduction in airspeed or increase in propeller load will cause the propeller blade angle to decrease.

The propeller's constant-speed range, defined by the high and low pitch stops, is the range of possible blade angles for a constant-speed propeller. As long as the propeller blade angle is within the constant-speed range and not against either pitch stop, a constant engine rpm will be maintained. If the propeller blades contact a pitch stop, the engine rpm will increase or decrease as appropriate, with changes in airspeed and propeller load. For example, once a specific rpm has been selected, if aircraft speed decreases enough to rotate the propeller blades until they contact the low pitch stop, any further decrease in airspeed will cause engine rpm to decrease the same way as if a fixed-pitch propeller were installed. The same holds true when an aircraft equipped with a constant-speed propeller accelerates to a faster airspeed. As the aircraft accelerates, the propeller blade angle increases to maintain the selected rpm until the high pitch stop is reached. Once this occurs, the blade angle cannot increase any further and engine rpm increases (Figure 4-6).

Figure 4-6 Engine power output is indicated on the manifold pressure gauge

On aircraft equipped with a constant-speed propeller, power output is controlled by the throttle and indicated by a manifold pressure gauge. The gauge measures the absolute pressure of the fuel/air mixture inside the intake manifold and is more correctly a measure of manifold absolute pressure (MAP). At a constant rpm and altitude, the amount of power produced is directly related to the fuel/air flow being delivered to the combustion chamber. As the throttle setting is increased, more fuel and air flows to the engine and MAP increases. When the engine is not running, the manifold pressure gauge indicates ambient air pressure (i.e., 29.92 inches mercury (29.92in Hg)). When the engine is started, the manifold pressure indication will decrease to a value less than ambient pressure (i.e., idle at 12in Hg). Engine failure or power loss is indicated on the manifold gauge as an increase in manifold pressure relative to the ambient air pressure at the altitude where the failure occurred.

The manifold pressure gauge is color coded to indicate the engine's operating range. The face of the manifold pressure gauge contains a green arc to show the normal operating range, and a red radial line to indicate the upper limit of manifold pressure.

For any given rpm, there is a manifold pressure that should not be exceeded. If manifold pressure is excessive for a given rpm, the pressure within the cylinders could be exceeded, placing undue stress on the cylinders. If repeated too frequently, this stress can weaken the cylinder components and eventually cause engine failure. As a general rule, manifold pressure (inches) should be less than the rpm.

A pilot can avoid conditions that overstress the cylinders by being constantly aware of the rpm, especially when increasing the manifold pressure. Pilots should conform to the manufacturer's recommendations for power settings of a particular engine to maintain the proper relationship between manifold pressure and rpm.

4.2 Induction Systems

The induction system brings in air from the outside, mixes it with fuel, and delivers the fuel/air mixture to the cylinder where combustion occurs. Outside air enters the induction system through an intake port on the front of the engine cowling. This port normally contains an air filter that inhibits the entry of dust and other foreign objects. Since the filter may occasionally become clogged, an alternate source of air must be available. Usually, the alternate air comes from inside the engine cowling, where it bypasses a clogged air filter. Some alternate air sources function automatically, while others operate manually.

Two types of induction systems are commonly used in small aircraft engines:

(1) The carburetor system, which mixes the fuel and air in the carburetor before this mixture enters the intake manifold.

(2) The fuel injection system, which mixes the fuel and air immediately before entry into each cylinder or injects fuel directly into each cylinder.

4.3 Fuel Systems

The fuel system is designed to provide an uninterrupted flow of clean fuel from the fuel tanks to the engine. The fuel must be available to the engine under all conditions of engine power, altitude, attitude, and during all approved flight maneuvers. Two common classifications apply to fuel systems in small aircraft: gravity-feed and fuel-pump systems.

The gravity-feed system utilizes the force of gravity to transfer the fuel from the tanks to the engine. For example, on high-wing airplanes, the fuel tanks are installed in the wings. This places the fuel tanks above the carburetor, and the fuel is gravity fed through the system and into the carburetor (Figure 4-7). If the design of the aircraft is such that gravity cannot be used to transfer fuel, fuel pumps are installed. For example, on low-wing airplanes, the fuel tanks in the wings are located below the carburetor (Figure 4-8).

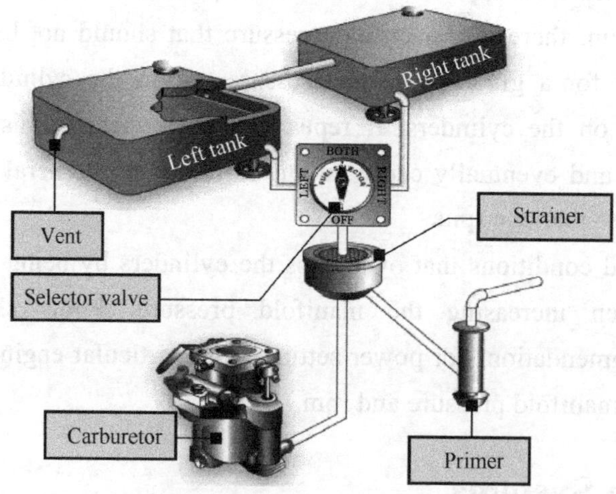

Figure 4-7　Gravity-feed system

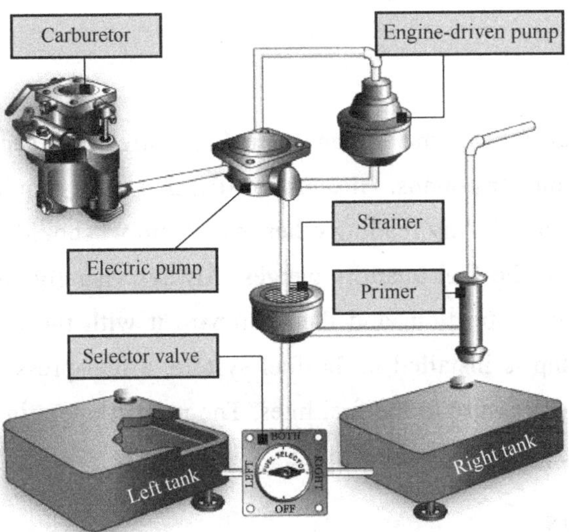

Figure 4-8 Fuel-pump system

Aircraft with fuel-pump systems have two fuel pumps. The main pump system is engine driven with an electrically driven auxiliary pump provided for use in engine starting and in the event the engine pump fails. The auxiliary pump, also known as a boost pump, provides added reliability to the fuel system. The electrically driven auxiliary pump is controlled by a switch in the flight deck(Figure 4-8).

4.3.1 Fuel Primer

Both gravity-feed and fuel-pump systems may incorporate a fuel primer into the system. The fuel primer is used to draw fuel from the tanks to vaporize fuel directly into the cylinders prior to starting the engine. During cold weather, when engines are difficult to start, the fuel primer helps because there is not enough heat available to vaporize the fuel in the carburetor. It is important to lock the primer in place when it is not in use. If the knob is free to move, it may vibrate out during flight and can cause an excessively rich mixture. To avoid overpriming, read the priming instructions for the aircraft.

4.3.2 Fuel Tanks

The fuel tanks, normally located inside the wings of an airplane, have a filler opening on top of the wing through which they can be filled. A filler cap covers this opening. The tanks are vented to the outside to maintain atmospheric pressure inside the tank. They may be vented through the filler cap or through a tube extending through the surface of the wing. Fuel tanks also include an overflow drain that may stand alone or be collocated with the fuel tank vent. This allows fuel to expand with increases in temperature without damage to the tank itself. If the tanks have been filled on a hot day, it is not unusual to see fuel coming from the overflow drain.

4.3.3 Fuel Gauges

The fuel quantity gauges indicate the amount of fuel measured by a sensing unit in each fuel tank and is displayed in gallons or pounds. Aircraft certification rules require accuracy in fuel gauges only when they read "empty". Any reading other than "empty" should be verified. Do not depend solely on the accuracy of the fuel quantity gauges. Always visually check the fuel level in each tank during the preflight inspection, and then compare it with the corresponding fuel quantity indication. If a fuel pump is installed in the fuel system, a fuel pressure gauge is also included. This gauge indicates the pressure in the fuel lines. The normal operating pressure can be found in the AFM/POH or on the gauge by color coding.

4.3.4 Fuel Selectors

The fuel selector valve allows selection of fuel from various tanks. A common type of selector valve contains four positions: LEFT, RIGHT, BOTH, and OFF. Selecting the LEFT or RIGHT position allows fuel to feed only from that tank, while selecting the BOTH position feeds fuel from both tanks. The LEFT or RIGHT position may be used to balance the amount of fuel remaining in each wing tank (Figure 4-9).

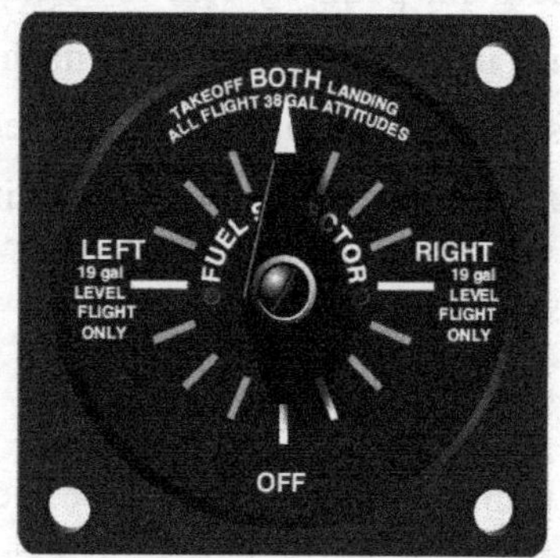

Figure 4-9 Fuel selector valve

Fuel placards will show any limitations on fuel tank usage, such as "level flight only" and/or "both" for landings and takeoffs.

Regardless of the type of fuel selector in use, fuel consumption should be monitored closely to ensure that a tank does not run completely out of fuel. Running a fuel tank dry will not only cause the engine to stop, but running for prolonged periods on one tank causes an unbalanced fuel load between tanks. Running a tank completely dry may allow air to enter the fuel system and cause vapor lock, which makes it difficult to restart the engine. On fuel-injected engines, the fuel becomes so hot it vaporizes in the fuel line, not allowing fuel to reach the cylinders.

4.3.5 Fuel Strainers, Sumps and Drains

After leaving the fuel tank and before entering the carburetor, the fuel passes through a strainer which removes any moisture and other sediments in the system. Since these contaminants are heavier than aviation fuel, they settle in a sump at the bottom of the strainer assembly. A sump is a low point in a fuel system and/or fuel tank. The fuel system may contain a sump, fuel strainer, and fuel tank drains, which may be collocated.

The fuel strainer should be drained before each flight. Fuel samples should be drained and checked visually for water and contaminants.

Water in the sump is hazardous because in cold weather the water can freeze and block fuel lines. In warm weather, it can flow into the carburetor and stop the engine. If water is present in the sump, more water in the fuel tanks is probable and they should be drained until there is no evidence of water. Never take off until all water and contaminants have been removed from the engine fuel system.

4.3.6 Fuel Grades

Aviation gasoline (AVGAS) is identified by an octane or performance number (grade), which designates the antiknock value or knock resistance of the fuel mixture in the engine cylinder. The higher the grade of gasoline, the more pressure the fuel can withstand without igniting. Lower grades of fuel are used in lower-compression engines because these fuels ignite at a lower temperature. Higher grades are used in higher-compression engines, because they ignite at higher temperatures, but not prematurely. If the proper grade of fuel is not available, use the next higher grade as a substitute. Never use a grade lower than recommended. This can cause the cylinder head temperature and engine oil temperature to exceed their normal operating ranges, which may result in unwanted ignition.

Several grades of AVGAS are available. Care must be exercised to ensure that the correct aviation grade is being used for the specific type of engine. The proper fuel grade is stated in the AFM/POH, on placards in the flight deck, and next to the filler caps. Auto gas should NEVER be used in aircraft engines unless the aircraft has been modified with a Supplemental Type Certificate (STC) issued by the Federal Aviation Administration (FAA).

The current method identifies AVGAS for aircraft with reciprocating engines by the octane and performance number, along with the abbreviation AVGAS. These aircraft use AVGAS 80, 100, and 100LL. Although AVGAS 100LL performs the same as grade 100, the "LL" indicates it has a low lead content. Fuel for aircraft with turbine engines is classified as JET A, JET A-1, and JET B. Jet fuel is basically kerosene and has a distinctive kerosene smell. Since use of the correct fuel is critical, dyes are added to help identify the type and grade of fuel (Figure 4-10).

In addition to the color of the fuel itself, the color-coding system extends to decals various airport fuel handling equipment. For example, all AVGAS is identified by name, using white letters on a red background. In contrast, turbine fuels are identified by white letters on a black background.

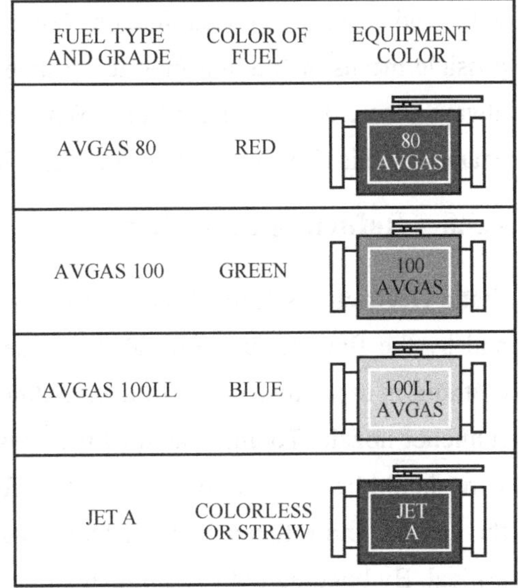

Figure 4-10 Aviation fuel color-coding system

4.3.7 Fuel Contamination

Accidents attributed to powerplant failure from fuel contamination have often been traced to:

(1) Inadequate preflight inspection by the pilot.

(2) Servicing aircraft with improperly filtered fuel from small tanks or drums.

(3) Storing aircraft with partially filled fuel tanks.

(4) Lack of proper maintenance.

Fuel should be drained from the fuel strainer quick drain and from each fuel tank sump into a transparent container, and then checked for dirt and water. When the fuel strainer is being drained, water in the tank may not appear until all the fuel has been drained from the lines leading to the tank. This indicates that water remains in the tank, and is not forcing the fuel out of the fuel lines leading to the fuel strainer. Therefore, drain enough fuel from the fuel strainer to be certain that fuel is being drained from the tank. The amount will depend on the length of fuel line from the tank to the drain. If water or other contaminants are found in the first sample, drain further samples until no trace appears.

Water may also remain in the fuel tanks after the drainage from the fuel strainer has ceased to show any trace of water. This residual water can be removed only by draining the fuel tank sump drains.

Water is the principal fuel contaminant. Suspended water droplets in the fuel can be identified by a cloudy appearance of the fuel, or by the clear separation of water from the colored fuel, which occurs after the water has settled to the bottom of the tank. As a safety measure, the fuel sumps should be drained before every flight during the preflight inspection.

Fuel tanks should be filled after each flight or after the last flight of the day to prevent moisture condensation within the tank. To prevent fuel contamination, avoid refueling from cans and drums.

In remote areas or in emergency situations, there may be no alternative to refueling from sources with inadequate anti-contamination systems. While a chamois skin and funnel may be the only possible means of filtering fuel, using them is hazardous. Remember, the use of a chamois will not always ensure decontaminated fuel. Worn-out chamois will not filter water; neither will a new, clean chamois that is already water-wet or damp. Most imitation chamois skins will not filter water.

4.3.8 Refueling Procedures

Static electricity is formed by the friction of air passing over the surfaces of an aircraft in flight and by the flow of fuel through the hose and nozzle during refueling. Nylon, Dacron, or wool clothing is especially prone to accumulate and discharge static electricity from the person to the funnel or nozzle. To guard against the possibility of static electricity igniting fuel fumes, a ground wire should be attached to the aircraft before the fuel cap is removed from the tank. Because both the aircraft and refueler have different static charges, bonding both components to each other is critical. By bonding both components to each other, the static differential charge is equalized. The refueling nozzle should be bonded to the aircraft before refueling begins and should remain bonded throughout the refueling process. When a fuel truck is used, it should be grounded prior to the fuel nozzle contacting the aircraft.

If fueling from drums or cans is necessary, proper bonding and grounding connections are important. Drums should be placed near grounding posts and the following sequence of connections is observed:

(1) Drum to ground.

(2) Ground to aircraft.

(3) Bond drum to aircraft.

(4) Nozzle to aircraft before the fuel cap is removed.

When disconnecting, reverse the order.

The passage of fuel through a chamois increases the charge of static electricity and the danger of sparks. The aircraft must be properly grounded and the nozzle, chamois filter, and funnel bonded to the aircraft. If a can is used, it should be connected to either the grounding post or the funnel. Under no circumstances should a plastic bucket or similar nonconductive container be used in this operation.

4.4 Starting Systems

Most small aircraft use a direct-cranking electric starter system. This system consists of a source of electricity, wiring, switches, and solenoids to operate the starter and a starter motor. Most aircraft have starters that automatically engage and disengage when operated, but some older aircraft have starters that are mechanically engaged by a lever actuated by the pilot. The starter engages the aircraft flywheel, rotating the engine at a speed that allows the engine to start and maintain operation.

Electrical power for starting is usually supplied by an onboard battery, but can also be supplied by an external power receptacle. When the battery switch is turned on, electricity is supplied to the main power bus bar through the battery solenoid. Both the starter and the starter switch draw current from the main bus bar, but the starter will not operate until the starting solenoid is energized by the starter switch being turned to the "start" position. When the starter switch is released from the "start" position, the solenoid removes power from the starter motor. The starter motor is protected from being driven by the engine through a clutch in the starter drive that allows the engine to run faster than the starter motor (Figure 4-11).

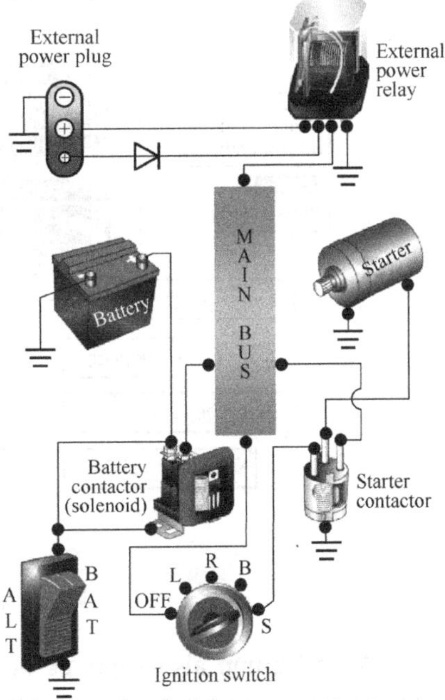

Figure 4-11 Typical starting circuit

When starting an engine, we should observe the rules of safety and courtesy. One of the most important is to make sure there is no one near the propeller. In addition, the wheels should be chocked and the brakes set, to avoid hazards caused by unintentional movement. To avoid damage to the propeller and property, the aircraft should be in an area where the propeller will not stir up gravel or dust.

4.5 Oil Systems

The engine oil system performs several important functions:

(1) Lubrication of the engine's moving parts.

(2) Cooling of the engine by reducing friction.

(3) Removing heat from the cylinders.

(4) Providing a seal between the cylinder walls and pistons.

(5) Carrying away contaminants.

Reciprocating engines use either a wet-sump or a dry-sump oil system. In a wet-sump system, the oil is located in a sump, which is an integral part of the engine. In a dry-sump system, the oil is contained in a separate tank, and circulated through the engine by pumps (Figure 4-12).

The main component of a wet-sump system is the oil pump, which draws oil from the sump and routes it to the engine. After the oil passes through the engine, it returns to the sump. In some engines, additional lubrication is supplied by the rotating crankshaft, which splashes oil onto portions of the engine.

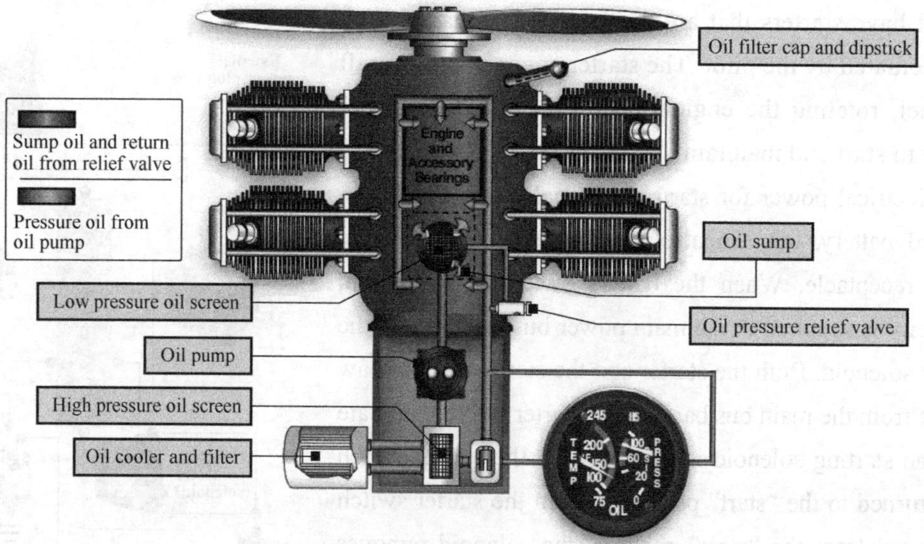

Figure 4-12　Wet-sump oil system

An oil pump also supplies oil pressure in a dry-sump system, but the source of the oil is located external to the engine in a separate oil tank. After oil is routed through the engine, it is pumped from the various locations in the engine back to the oil tank by scavenge pumps. Dry-sump systems allow for a greater volume of oil to be supplied to the engine, which makes them more suitable for very large reciprocating engines.

The oil pressure gauge provides a direct indication of the oil system operation. It ensures the pressure in pounds per square inch (psi) of the oil supplied to the engine. Green indicates the normal operating range, while red indicates the minimum and maximum pressures. There should be an indication of oil pressure during engine start. Refer to the AFM/POH for manufacturer limitations.

The oil temperature gauge measures the temperature of oil. A green area shows the normal operating range and the red line indicates the maximum allowable temperature. Unlike oil pressure, changes in oil temperature occur more slowly. This is particularly noticeable after starting a cold engine, when it may take several minutes or longer for the gauge to show any increase in oil temperature.

Check oil temperature periodically during flight especially when operating in high or low ambient air temperature. High oil temperature indications may signal a plugged oil line, a low oil quantity, a blocked oil cooler, or a defective temperature gauge. Low oil temperature indications may signal improper oil viscosity during cold weather operations.

The oil filler cap and dipstick (for measuring the oil quantity) are usually accessible through a panel in the engine cowling. If the quantity does not meet the manufacturer's recommended operating levels, oil should be added. The AFM/POH or placards near the access panel provide information about the correct oil type and weight, as well as the minimum and maximum oil quantity (Figure 4-13).

Figure 4-13 Always check the engine oil level during the preflight inspection

4.6 Cooling Systems

The burning fuel within the cylinders produces intense heat, most of which is expelled through the exhaust system. Much of the remaining heat, however, must be removed, or at least dissipated, to prevent the engine from overheating. Otherwise, the extremely high engine temperatures can lead to loss of power, excessive oil consumption, detonation, and serious engine damage.

While the oil system is vital to the internal cooling of the engine, an additional method of cooling is necessary for the engine's external surface. Most small aircraft are air cooled, although some are liquid cooled.

Air cooling is accomplished by air flowing into the engine compartment through openings in front of the engine cowling. Baffles route this air over fins attached to the engine cylinders, and other parts of the engine, where the air absorbs the engine heat. Expulsion of the hot air takes place through one or more openings in the lower, aft portion of the engine cowling (Figure 4-14).

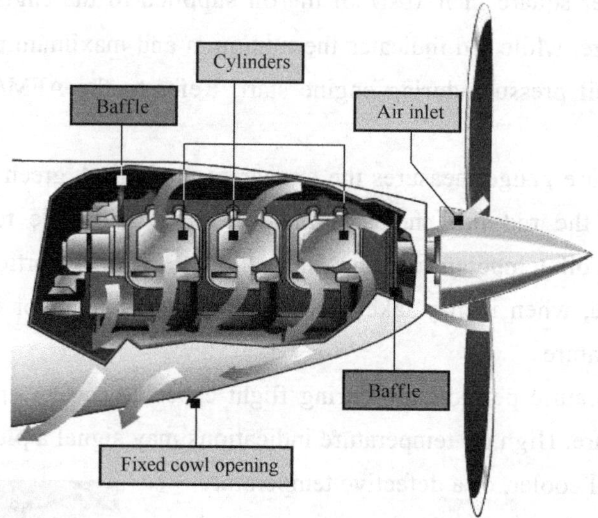

Figure 4-14　Outside air aids in cooling the engine

The outside air enters the engine compartment through an inlet behind the propeller hub. Baffles direct it to the hottest parts of the engine, primarily the cylinders, which have fins that increase the area exposed to the airflow.

The air cooling system is less effective during ground operations, takeoffs, go-arounds, and other periods of high-power, low-airspeed operation. Conversely, high-speed descents provide excess air and can cool the engine, subjecting it to abrupt temperature fluctuations.

Operating the engine at higher than its designed temperature can cause loss of power, excessive oil consumption, and detonation. It can also lead to serious permanent damage, such as scoring the cylinder walls, damaging the pistons and rings, and burning and warping the valves. Monitoring the flight deck engine temperature instruments will aid in avoiding high operating temperature.

Under normal operating conditions in aircraft not equipped with cowl flaps, the engine temperature can be controlled by changing the airspeed or the power output of the engine. High engine temperatures can be decreased by increasing the airspeed and/or reducing the power.

The oil temperature gauge gives an indirect and delayed indication of rising engine temperature, but can be used for determining engine temperature if this is the only means available.

Most aircraft are equipped with a cylinder-head temperature gauge which indicates a direct and immediate cylinder temperature change. This instrument is calibrated in degrees Celsius or Fahrenheit, and is usually color coded with a green arc to indicate the normal operating range. A red line on the instrument indicates maximum allowable cylinder head temperature.

To avoid excessive cylinder head temperatures, increase airspeed, enrich the mixture, and/or reduce power. Any of these procedures can help to reduce the engine temperature. On aircraft equipped with cowl flaps, use the cowl flap positions to control the temperature. Cowl flaps are hinged covers that fit over the opening through which the hot air is expelled. If the engine temperature is low, the cowl flaps can be closed, thereby restricting the flow of expelled hot air and increasing engine temperature. If the engine temperature is high, the cowl flaps can be opened to permit a greater flow of air through the system, thereby decreasing the engine temperature.

4.7 Exhaust Systems

Engine exhaust systems vent the burned combustion gases overboard, provide heat for the cabin, and defrost the windscreen. An exhaust system has exhaust piping attached to the cylinders, as well as a muffler and a muffler shroud. The exhaust gases are pushed out of the cylinder through the exhaust valve and then through the exhaust pipe system to the atmosphere.

For cabin heat, outside air is drawn into the air inlet and is ducted through a shroud around the muffler. The muffler is heated by the exiting exhaust gases and, in turn, heats the air around the muffler. This heated air is then ducted to the cabin for heat and defrosting applications. The heat and defrosting are controlled in the flight deck, and can be adjusted to the desired level.

Exhaust gases contain large amounts of carbon monoxide, which is odorless and colorless. Carbon monoxide is deadly, and its presence is virtually impossible to detect. The exhaust system must be in good condition and free of cracks.

Some exhaust systems have an EGT probe. This probe transmits the EGT to an instrument in the flight deck. The EGT gauge measures the temperature of the gases at the exhaust manifold. This temperature varies with the ratio of fuel to air entering the cylinders and can be used as a basis for regulating the fuel/air mixture. The EGT gauge is highly accurate in indicating the correct mixture setting. When we use the EGT to aid in leaning the fuel/air mixture, fuel consumption can be reduced. For specific procedures, refer to the manufacturer's recommendations for leaning the mixture.

4.8 Electrical System

Most aircraft are equipped with either a 14 or a 28 volt direct current electrical system. A basic aircraft electrical system consists of the following components:

(1) Alternator/generator;

(2) Battery;

(3) Master/battery switch;

(4) Alternator/generator switch;

(5) Bus bar, fuses, and circuit breakers;

(6) Voltage regulator;

(7) Ammeter/loadmeter;

(8) Associated electrical wiring.

Engine-driven alternators or generators supply electric current to the electrical system. They also maintain a sufficient electrical charge in the battery. Electrical energy stored in a battery provides a source of electrical power for starting the engine and a limited supply of electrical power for use in the event that the alternator or generator fails.

Most direct-current generators will not produce a sufficient amount of electrical current at low engine rpm to operate the entire electrical system. During operations at low engine rpm, the electrical needs must be drawn from the battery, which can quickly be depleted.

Alternators have several advantages over generators. Alternators produce sufficient current to operate the entire electrical system, even at slower engine speeds, by producing alternating current, which is converted to direct current. The electrical output of an alternator is more constant throughout a wide range of engine speeds.

Some aircraft have receptacles to which an external ground power unit (GPU) may be connected to provide electrical energy for starting. They are very useful, especially during cold weather starting. Follow the manufacturer's recommendations for engine starting using a GPU.

The electrical system is turned on or off with a master switch. Turning the master switch to the ON position provides electrical energy to all the electrical equipment circuits except the ignition system. Equipment that commonly uses the electrical system for its source of energy includes:

(1) Position lights;

(2) Anticollision lights;

(3) Landing lights;

(4) Taxi lights;

(5) Interior cabin lights;

(6) Instrument lights;

(7) Radio equipment;

(8) Turn indicator;

(9) Fuel gauges;

(10) Electric fuel pump;

(11) Stall warning system;

(12) Pitot heat;

(13) Starting motor.

Many aircraft are equipped with a battery switch that controls the electrical power to the aircraft in a manner similar to the master switch. In addition, an alternator switch is installed which permits the pilot to exclude the alternator from the electrical system in the event of alternator failure (Figure 4-15).

With the alternator half of the switch in the OFF position, the entire electrical load is placed on the battery. All nonessential electrical equipment should be turned off to conserve battery power. A bus bar is used as a terminal in the aircraft electrical system to connect the main electrical system to the equipment using electricity as a source of power. This simplifies the wiring system and provides a common point from which voltage can be distributed throughout the system. Fuses or circuit breakers are used in the electrical system to protect the circuits and equipment from electrical overload.

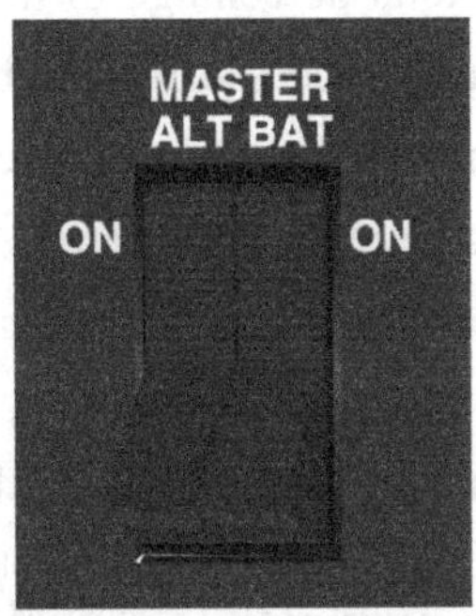

Figure 4-15 On this master switch, the left half is for the alternator and the right half is for the battery

Spare fuses of the proper amperage limit should be carried in the aircraft to replace defective or blown fuses. Circuit breakers have the same function as a fuse but can be manually reset, rather than replaced, if an overload condition occurs in the electrical system. Placards at the fuse or circuit breaker panel identify the circuit by name and show the amperage limit.

An ammeter is used to monitor the performance of the aircraft electrical system. The ammeter shows if the alternator/generator is producing an adequate supply of electrical power. It also indicates whether or not the battery is receiving an electrical charge.

Ammeters are designed with the zero point in the center of the face and a negative or positive indication on either side (Figure 4-16). When the pointer of the ammeter is on the plus side, it shows the charging rate of the battery. A minus indication means more current is being drawn from the battery than is being replaced. A full-scale minus deflection indicates a malfunction of the alternator/generator. A full-scale positive deflection indicates a malfunction of the regulator. In either case, consult the AFM or POH for appropriate action to be taken.

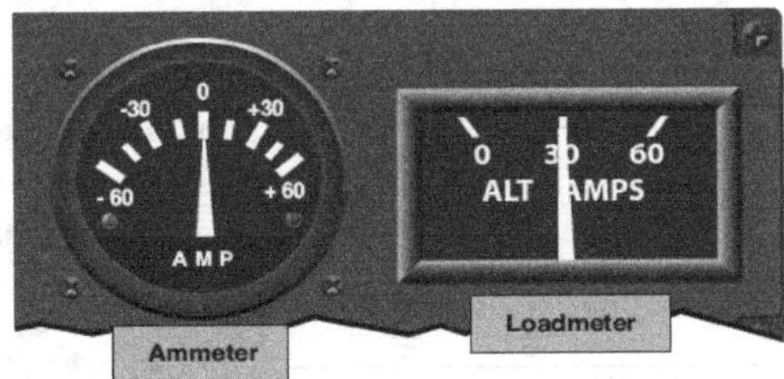

Figure 4-16 Ammeter and loadmeter

Not all aircraft are equipped with an ammeter. Some have a warning light that, when lighted, indicates a discharge in the system as a generator/alternator malfunction. Refer to the AFM or POH for appropriate action to be taken (Figure 4-17).

Another electrical monitoring indicator is a loadmeter. This type of gauge has a scale beginning with zero and shows the load being placed on the alternator/generator. The loadmeter reflects the total percentage of the load placed on the generating capacity of the electrical system by the electrical accessories and battery. When all electrical components are turned off, it reflects only the amount of charging current demanded by the battery.

A voltage regulator controls the rate of charge to the battery by stabilizing the generator or alternator electrical output. The generator/alternator voltage output should be higher than that of the battery voltage. For example, a 12-volt battery would be fed by a generator/alternator system of approximately 14 volts. The difference in voltage keeps the battery charged(Figure 4-16).

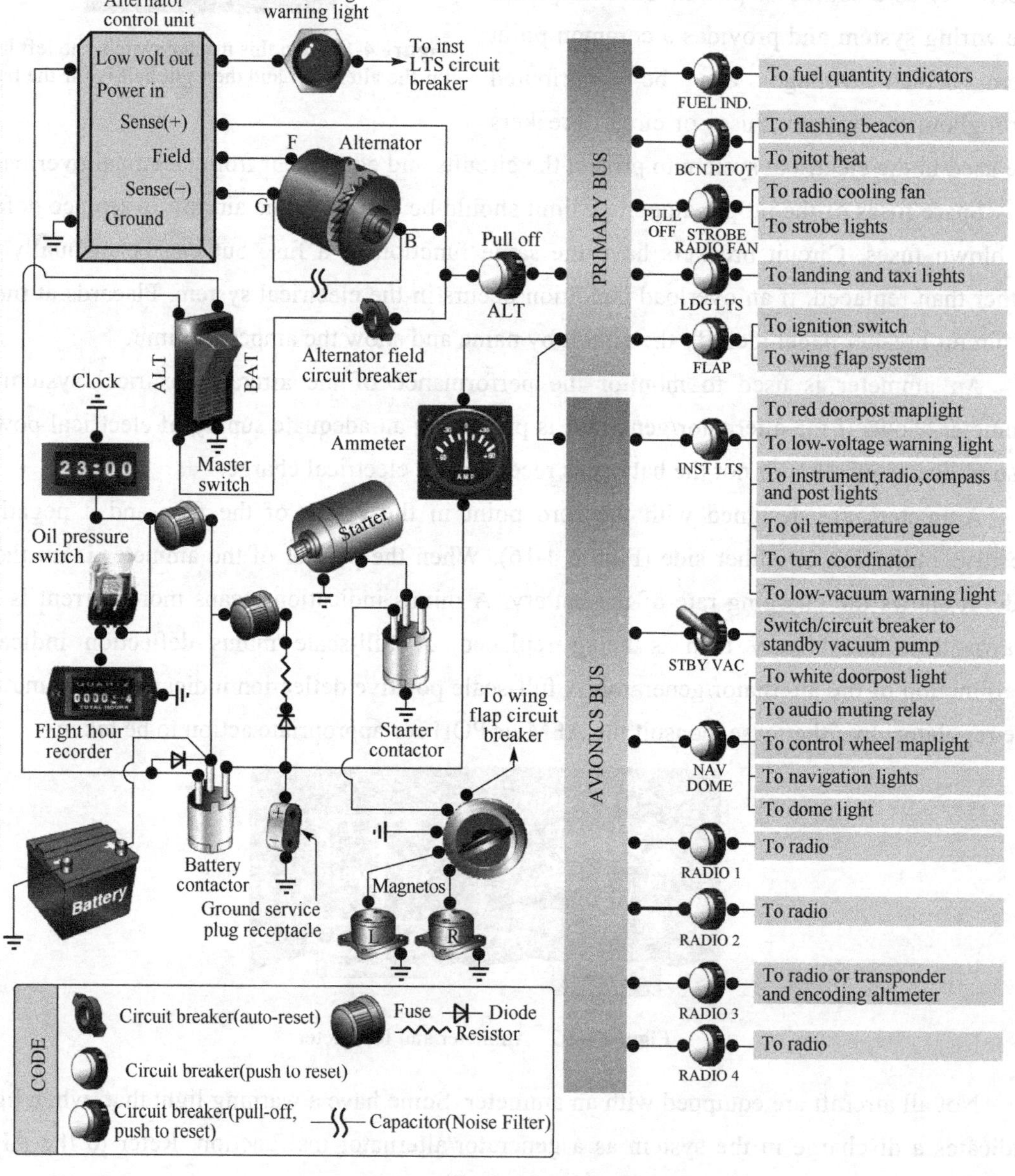

Figure 4-17　Electrical system schematic

4.9 Hydraulic Systems

There are multiple applications for hydraulic use in aircraft, depending on the complexity of the aircraft. For example, hydraulics is often used on small airplanes to operate wheel brakes, retractable landing gear, and some constant-speed propellers. On large airplanes, hydraulics is used for flight control surfaces, wing flaps, spoilers, and other systems.

A basic hydraulic system consists of a reservoir, pump (either hand, electric, or engine driven), a filter to keep the fluid clean, selector valve to control the direction of flow, relief valve to relieve excess pressure, and an actuator (Figure 4-18).

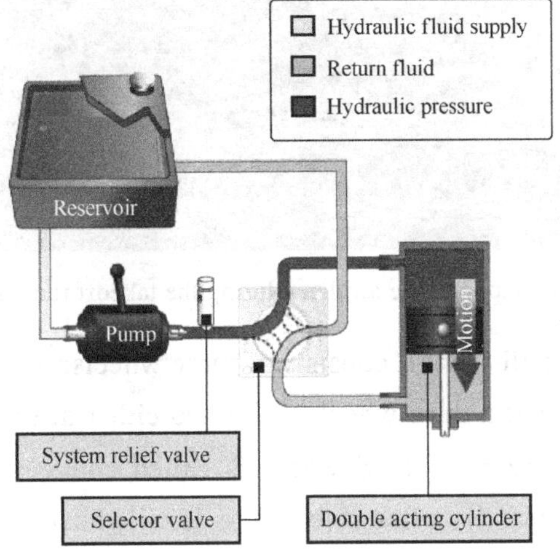

Figure 4-18 Basic hydraulic system

The hydraulic fluid is pumped through the system to an actuator or servo. A servo is a cylinder with a piston inside that turns fluid power into work and creates the power needed to move an aircraft system or flight control. Servos can be either single-acting or double-acting, based on the needs of the system. This means that the fluid can be applied to one or both sides of the servo, depending on the servo type. A single-acting servo provides power in one direction. The selector valve allows the fluid direction to be controlled. This is necessary for operations such as the extension and retraction of landing gear during which the fluid must work in two different directions. The relief valve provides an outlet for the system in the event of excessive fluid pressure in the system. Each system incorporates different components to meet the individual needs of different aircraft.

A mineral-based hydraulic fluid is the most widely used type for small aircraft. This type of hydraulic fluid, a kerosene-like petroleum product, has good lubricating properties, as well as additives to inhibit foaming and prevent the formation of corrosion. It is chemically stable, has very little viscosity change with temperature, and is dyed for identification. Since several types of hydraulic fluids are commonly used, an aircraft must be serviced with the type specified by the manufacturer. Refer to the AFM/POH or the Maintenance Manual.

4.9.1 Landing Gear

The landing gear forms the principal support of an aircraft on the ground. The most common type of landing gear consists of wheels, but aircraft can also be equipped with floats for water operations or skis for landing on snow (Figure 4-19).

Figure 4-19 The landing gear supports the airplane during the takeoff run, landing, taxiing, and when parked

The landing gear on small aircraft consists of three wheels: two main wheels (one located on each side of the fuselage) and a third wheel positioned either at the front or rear of the airplane. Landing gear employing a rear-mounted wheel is called conventional landing gear. Airplanes with conventional landing gear are often referred to as tailwheel airplanes. When the third wheel is located on the nose, it is called a nosewheel, and the design is referred to as a tricycle gear. A steerable nosewheel or tailwheel permits the airplane to be controlled throughout all operations while on the ground.

4.9.2 Tricycle Landing Gear Airplanes

A tricycle gear airplane has three advantages:

(1) It allows more forceful application of the brakes during landings at high speeds without causing the aircraft to nose over.

(2) It permits better forward visibility for the pilot during takeoff, landing, and taxiing.

(3) It tends to prevent ground looping (swerving) by providing more directional stability during ground operation since the aircraft's center of gravity (CG) is forward of the main wheels. The forward CG keeps the airplane moving forward in a straight line rather than ground looping.

Nosewheels are either steerable or castering. Steerable nosewheels are linked to the rudders by cables or rods, while castering nosewheels are free to swivel. In both cases, the aircraft is steered using the rudder pedals. Aircraft with a castering nosewheel may require the pilot to combine the use of the rudder pedals with independent use of the brakes.

4.9.3 Tailwheel Landing Gear Airplanes

Tailwheel landing gear aircraft have two main wheels attached to the airframe ahead of its CG that support most of the weight of the structure. A tailwheel at the very back of the fuselage provides a third point of support. This type of installation was used early on large propeller aircraft and is suitable for venues where the ground is not solid enough. (Figure 4-20).

Figure 4-20 Tailwheel landing gear

With the CG located behind the main gear, directional control of this type aircraft becomes more difficult while on the ground. This is the main disadvantage of the tailwheel landing gear. For example, if the pilot allows the aircraft to swerve while aircraft roll on the ground at a low speed, he or she may not have sufficient rudder control and the CG will attempt to get ahead of the main gear which may cause the airplane to ground loop.

4.9.4 Fixed and Retractable Landing Gear

Landing gear can also be classified as either fixed or retractable. A fixed gear always remains extended and has the advantage of simplicity combined with low maintenance. A retractable gear is designed to streamline the airplane by allowing the landing gear to be stowed inside the structure during cruising flight (Figure 4-21).

(a) fixed

(b) retractable

Figure 4-21 Fixed and retractable gear airplanes

4.9.5 Brakes

Airplane brakes are located on the main wheels and are applied by either a hand control or by foot pedals (toe or heel). Foot pedals operate independently and allow for differential braking. During ground operations, differential braking can supplement nosewheel/tailwheel steering.

4.10 Oxygen Systems

Most high altitude aircraft come equipped with some type of fixed oxygen installation. If the aircraft does not have a fixed installation, portable oxygen equipment must be readily accessible during flight. The portable equipment usually consists of a container, regulator, mask outlet, and pressure gauge. Aircraft oxygen is usually stored in high pressure system containers of 1800–2200 psi. When the ambient temperature surrounding an oxygen cylinder decreases, pressure within that cylinder decreases because pressure varies directly with temperature if the volume of a gas remains constant. If a drop in indicated pressure on a supplemental oxygen cylinder is noted, there is no reason to suspect depletion of the oxygen supply, which has simply been compacted due to storage of the containers in an unheated area of the aircraft. High pressure oxygen containers should be marked with the psi tolerance (i.e., 1800 psi) before filling the container to that pressure. The containers should be supplied with aviation oxygen only, which is at least 99.5 percent pure oxygen. Industrial oxygen is not intended for breathing and may contain impurities, and medical oxygen contains water vapor that can freeze in the regulator when exposed to cold temperatures. To assure safety, periodic inspection and servicing of the oxygen system should be done.

An oxygen system consists of a mask or cannula and a regulator that supplies a flow of oxygen dependent upon cabin altitude. Cannulas are not approved for flights above 18000ft. Regulators approved for use up to 40000ft are designed to provide zero percent cylinder oxygen and 100 percent cabin air at cabin altitudes of 8000ft or less, with the ratio changing to 100 percent oxygen and zero percent cabin air at approximately 34000ft cabin altitude. Regulators approved up to 45000ft are designed to provide 40 percent cylinder oxygen and 60 percent cabin air at lower altitudes, with the ratio changing to 100 percent at the higher altitude. Pilots should avoid flying above 10000ft without oxygen during the day and above 8000ft at night (Figure 4-22).

Pilots should be aware of the danger of fire when using oxygen. Materials that are nearly fireproof in ordinary air may be susceptible to combustion in oxygen. Oils and greases may ignite if exposed to oxygen, and cannot be used for sealing the valves and fittings of oxygen equipment. Smoking during any kind of oxygen equipment use is prohibited. Before each flight, the pilot should thoroughly inspect and test all oxygen equipment. The inspection should include a thorough examination of the aircraft oxygen equipment, including available supply, an operational check of the system, and assurance that the supplemental oxygen is readily accessible.

The inspection should be accomplished with clean hands and should include a visual inspection of the mask and tubing for tears, cracks, or deterioration; the regulator for valve and lever condition and positions; oxygen quantity; and the location and functioning of oxygen pressure gauges, flow indicators and connections. The mask should be donned and the system should be tested. After any oxygen use, verify that all components and valves are shut off.

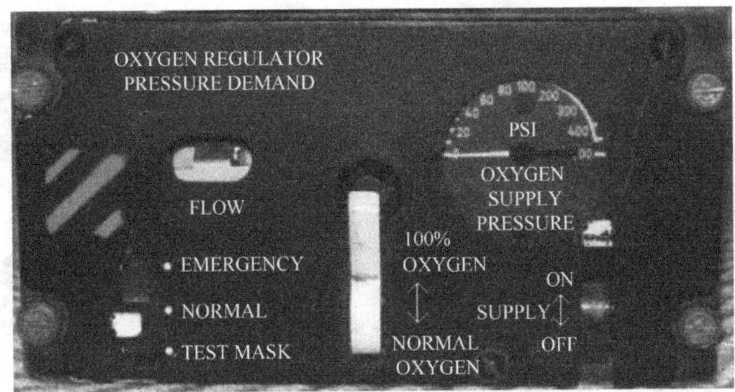

Figure 4-22 Oxygen system regulator

4.11 Anti-Ice Systems

Anti-icing equipment is designed to prevent the formation of ice, while deicing equipment is designed to remove ice once it has formed. These systems protect the leading edge of wing and tail surfaces, Pitot and static port openings, fuel tank vents, stall warning devices, windshields, and propeller blades. Ice detection lighting may also be installed on some aircraft to determine the extent of structural icing during night flights.

4.11.1 Airfoil Anti-Ice and Deicing

Inflatable deicing boots consist of a rubber sheet bonded to the leading edge of the airfoil. When ice builds up on the leading edge, an engine-driven pneumatic pump inflates the rubber boots. Many turboprop aircraft divert engine bleed air to the wing to inflate the rubber boots. Upon inflation, the ice is cracked and should fall off the leading edge of the wing. Deicing boots are controlled from the flight deck by a switch and can be operated in a single cycle or allowed to cycle at automatic, timed intervals (Figure 4-23).

Many deicing boot systems use the instrument system suction gauge and a pneumatic pressure gauge to indicate proper boot operation. These gauges have range markings that indicate the operating limits for boot operation. Some systems may also incorporate an annunciator light to indicate proper boot operation.

Proper maintenance and care of deicing boots are important for continued operation of this system. They need to be carefully inspected during preflight.

Another type of leading edge protection is the thermal anti-ice system. Heat provides one of the most effective methods for preventing ice accumulation on an airfoil. High performance turbine aircraft often direct hot air from the compressor section of the engine to the leading edge surfaces. The hot air heats the leading edge surfaces sufficiently to prevent the formation of ice.

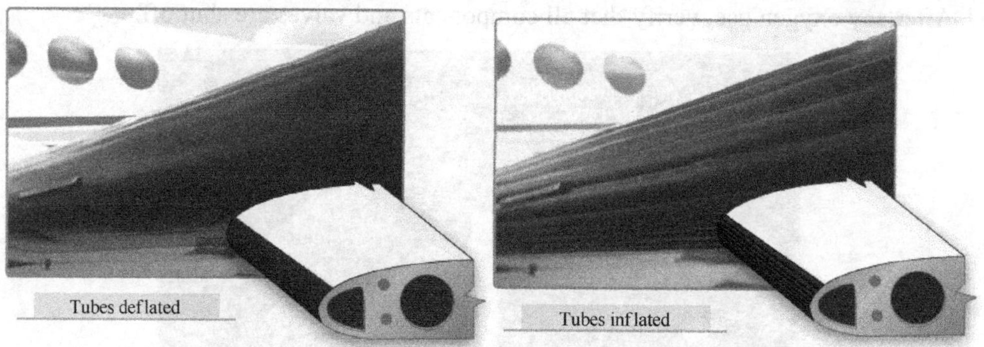

Figure 4-23 Deicing boots on the leading edge of the wing

An alternate type of leading edge protection that is not as common as thermal anti-ice and deicing boots is known as a weeping wing. The weeping-wing design uses small holes located in the leading edge of the wing to prevent the formation and build-up of ice. An antifreeze solution is pumped to the leading edge and weeps out through the holes.

4.11.2 Windscreen Anti-Ice

There are two main types of windscreen anti-ice systems. The first system directs a flow of alcohol to the windscreen. If used early enough, the alcohol will prevent ice from building up on the windscreen. The rate of alcohol flow can be controlled by a dial in the flight deck according to procedures recommended by the aircraft manufacturer.

Another effective method of anti-icing equipment is the electric heating method. Small wires or other conductive material is imbedded in the windscreen. The heater can be turned on by a switch in the flight deck, causing an electrical current to be passed across the shield through the wires to provide sufficient heat to prevent the formation of ice on the windscreen. The heated windscreen should only be used during flight. Do not leave it on during ground operations, as it can overheat and cause damage to the windscreen. Warning: the electrical current can cause compass deviation errors by as much as 40°.

4.11.3 Propeller Anti-Ice

Propellers are protected from icing by the use of alcohol or electrically heated elements. Some propellers are equipped with a discharge nozzle that is pointed toward the root of the blade. Alcohol is discharged from the nozzles, and centrifugal force drives the alcohol down the leading edge of the blade. The boots are also grooved to help direct the flow of alcohol. This prevents ice from forming on the leading edge of the propeller. Propellers can also be fitted with propeller anti-ice boots. The propeller boot is divided into two sections—the inboard and the outboard sections. The boots are imbedded with electrical wires that carry current for heating the propeller. The prop anti-ice system can be monitored for proper operation by monitoring the prop anti-ice ammeter. During the preflight inspection, check the propeller boots for proper operation. If a boot fails to heat one blade, an unequal blade loading can result, and may cause severe propeller vibration.

4.11.4 Other Anti-Ice and Deicing Systems

Pitot and static ports, fuel vents, stall-warning sensors, and other optional equipment may be heated by electrical elements. Operational checks of the electrically heated systems are to be checked in accordance with the AFM /POH.

Operation of aircraft anti-icing and deicing systems should be checked prior to encountering icing conditions. Encounters with structural ice require immediate action. Anti-icing and deicing equipment are not intended to sustain long-term flight in icing conditions.

Words and Expressions

1. Hydraulic System 液压系统

There are three types of hydraulic fluids currently being used in civil aircraft. They are:

vegetable base hydraulic fluid 植物液压油

mineral base hydraulic fluid 矿物液压油

phosphate ester base fluids 磷酸酯液压油

A basic hydraulic system includes (Figure 4-24):

① reservoir 液压油箱

② power-driven pump 动力驱动泵

③ filter 过滤器

④ pressure regulator 压力调节器

⑤ accumulator 蓄压器

⑥ check valve 单向阀

⑦ hand pump 手动泵

⑧ pressure gage 压力表

⑨ relief valve 减压阀，安全阀

⑩ actuating unit 动力传动装置

⑪ cylinder 汽缸

⑫ piston 活塞

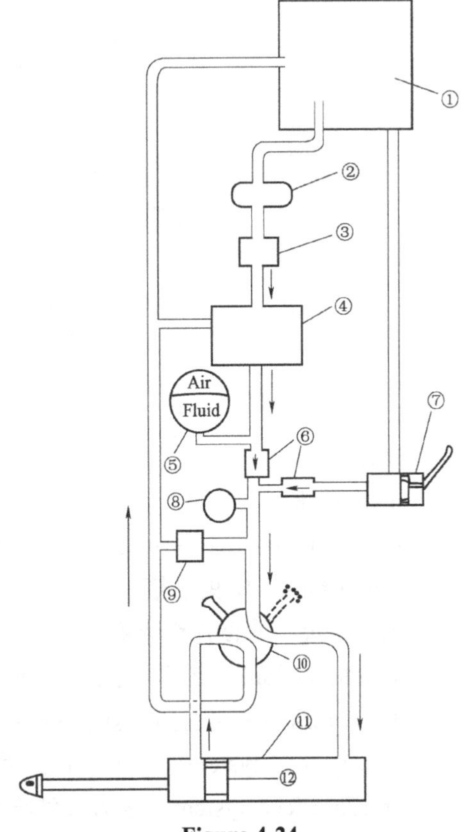

Figure 4-24

2.　Fuel System　燃油系统

aviation fuel　航空燃料

gasoline　汽油

kerosene　煤油

Two types of jet turbine engine fuel in common use today are:

jet turbine　喷气涡轮

kerosene grade turbine fuel　煤油涡轮燃料

a blend of gasoline and kerosene fractions　汽煤混合馏分燃料

Figure 4-25 shows the typical arrangement of the integral fuel tanks in a jet airliner. It has:

integral tank　整体油箱

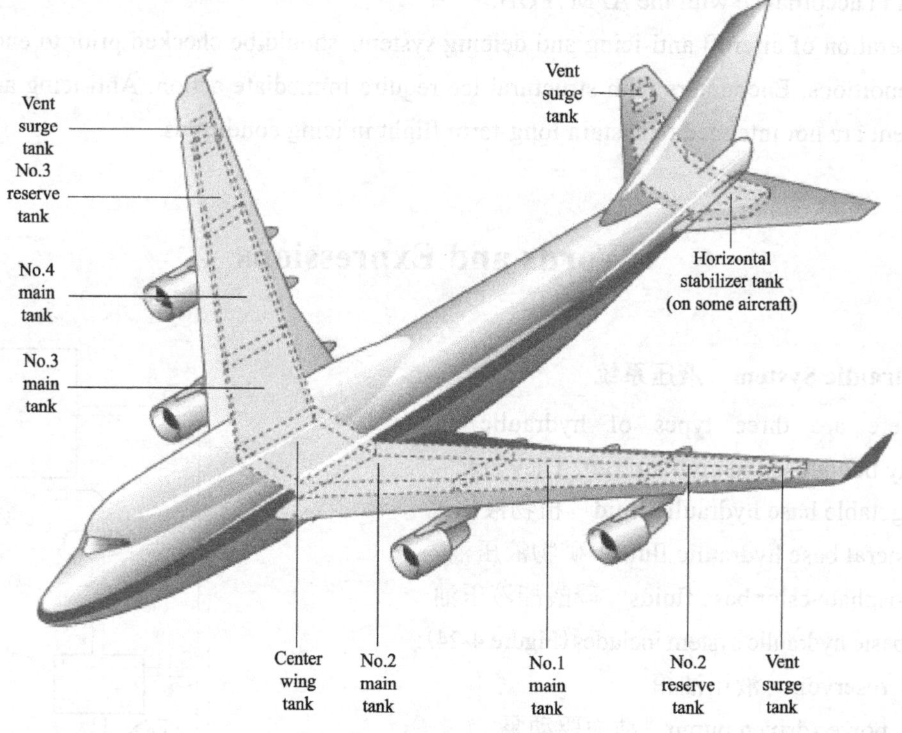

Figure 4-25

fuel tank　油箱

main tank　主油箱

reserve tank / auxiliary tank　储备油箱

center tank　中央油箱

trim tank　重心调整油箱

vent surge tank　通风(防震动)油箱

Two types of fuel feed system are used to supply the fuel to the engines. They are:

gravity feed system　重力式供给系统

pressure feed system　压力式供给系统

Figure 4-26 shows a typical four-engine manifold cross-feed system.

manifold cross-feed system 总管交输系统

engine 发动机

firewall shut-off valve 防火墙关断阀门

tank 油箱

filter 过滤器

manifold valve 总管阀

auxiliary tank 储备油箱

fuel manifold 燃油总管

refueling receptacle 燃油加注接头

booster / boost pump 增压泵

tank shut-off valve 油箱关断阀门

engine-driven fuel pump 发动机驱动燃油泵

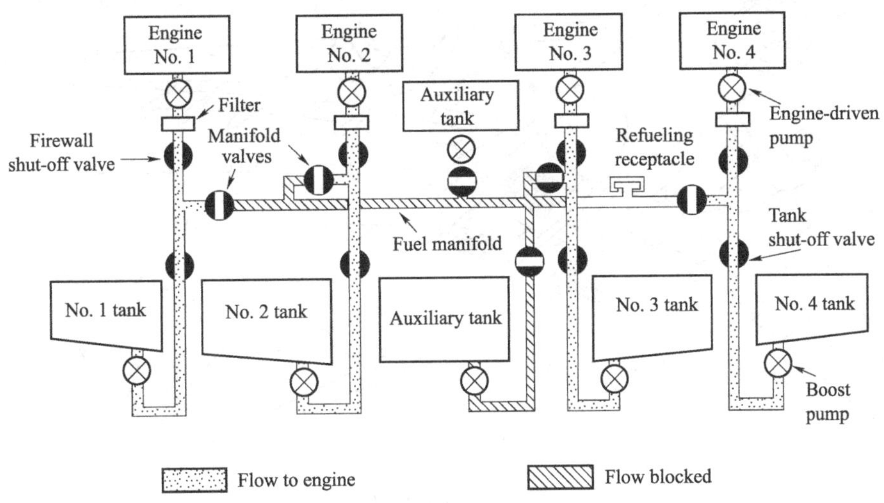

Figure 4-26

3. Landing Gear System 起落架系统

inboard retracting main gear 向内收起式主起落架

forward retracting nose gear 向前收起式前起落架

tail wheel gear 后三点式起落架

tricycle gear 前三点式起落架

landing gear bay 起落架舱

landing strut 起落架支柱

gear retraction 收起起落架

hand crank 手动式摇柄

nose wheel steering 前轮转向操纵

downlock actuator 下位锁作动器

downlock springs 下位锁弹簧

shock absorber　减震器
safety pin　安全销
lock stay　锁紧撑杆
retraction actuator　收起作动器
nose landing gear / NLG　前起落架(Figure 4-27)
main landing gear / MLG　主起落架(Figure 4-27)
actuating cylinder　作动气缸
downlock cylinder　下位锁气缸
landing gear selector valve　起落架换向阀
retracting cylinder　收起气缸

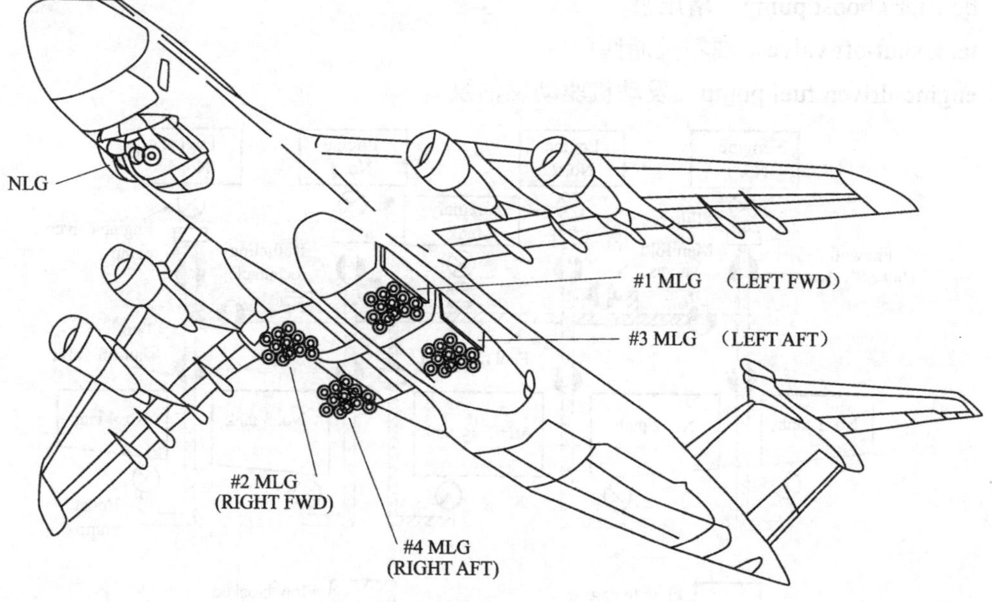

Figure 4-27

hydraulic landing gear retraction system　液压式起落架收起系统(Figure 4-28)
selector valve　选择阀/换向阀
orifice check valve　小孔单向阀
gear (actuating) cylinder　起落架作动气缸
sequence check valve (A, B, C, D)　序列单向阀
up-lock　上位锁(Figure 4-29)
down-lock　下位锁
door (actuating) cylinder　舱盖作动气缸
nose gear (actuating) cylinder　前起落架作动气缸
multiple-disk rotor type brakes　多盘转子型刹车(Figure 4-30)
pressure plate　压力板
rotor　转子

stator　定子

piston　活塞

adjuster　调节器

brake mount flange　刹车架凸缘

wear indicator　磨耗指示器

brake hose connection / hydraulic bleed port　刹车软管接头/液压放气口

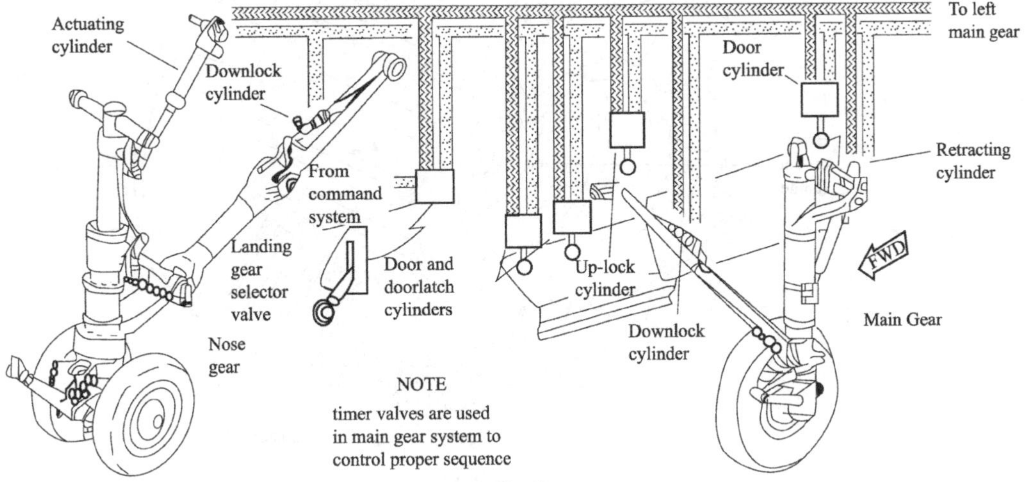

Figure 4-28

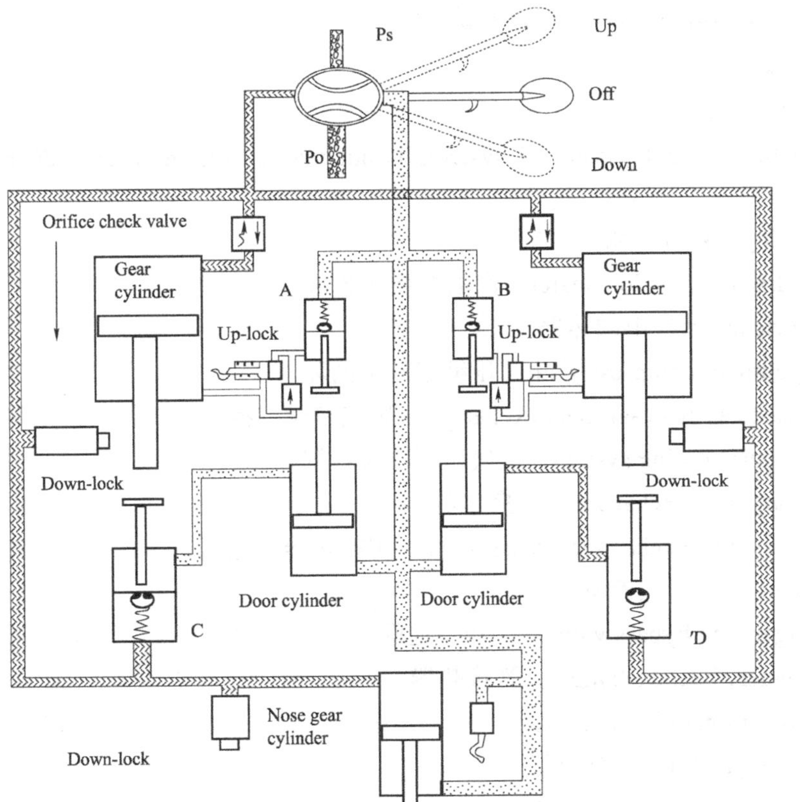

Figure 4-29

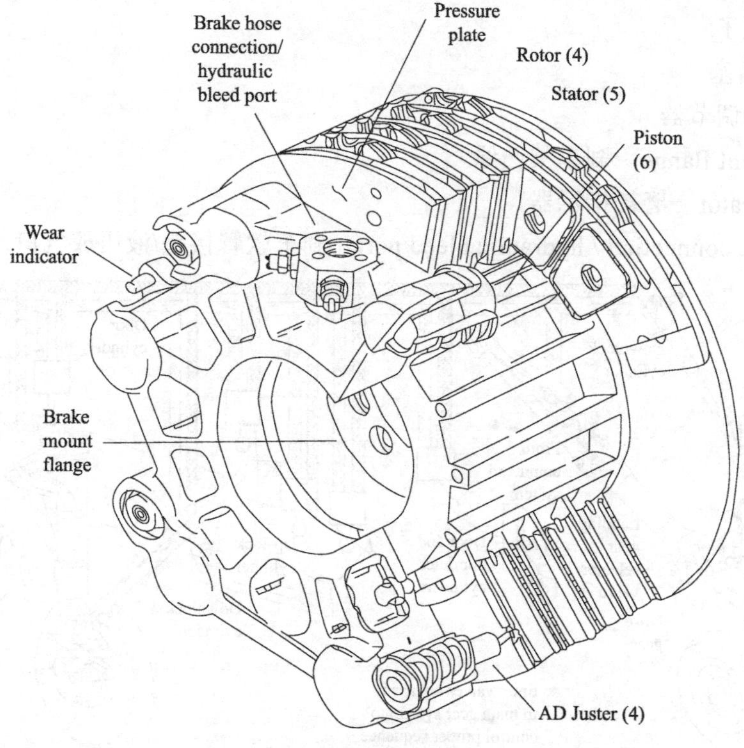

Figure 4-30

4. Air Conditioning System　空调系统

pressurization　增压

ventilation　通风

One of the following types of systems is put into operation when a degree of heating is needed.

(1) heating system　供热系统

gasoline combustion heater　汽油燃烧加热器

electric heater　电加热器

recycling of compressed air　压缩空气再循环

exhaust gas air-to-air heat exchanger　废气空对空换热器

(2) air cycle cooling system　空气循环冷却系统

primary heat exchanger　主换热器

primary heat exchanger bypass valve　主换热器旁通阀

shutoff valve　关断阀门

refrigeration bypass valve　制冷旁通阀

secondary heat exchanger　次换热器

refrigeration unit　制冷机组

water separator　水分离器

ram air valve　冲压空气阀

(3) vapor cycle cooling system (freon)　蒸汽循环冷却系统

evaporator　蒸发器

compressor　压缩器

condenser　冷凝器

expansion valve　膨胀阀

condenser fan　冷凝器风扇

receiver / freon storage　接收器

dryer　干燥器

surge valve　补偿阀

temperature control　温度控制

(4) cabin pressure control system　舱压控制系统

(5) cabin pressure regulator　座舱压力调节器

(6) outflow valve　排气阀门

(7) safety valve　安全阀

5. Oxygen System　供氧系统

There are three forms of oxygen on aircraft:

gaseous oxygen　气态氧

liquid oxygen　液态氧

chemical oxygen / solid oxygen　化学氧/固态氧

continuous flow system　恒量供氧式呼吸器

pressure demand system　压力需求式呼吸器

portable oxygen equipment　便携式供氧设备

The components of portable oxygen equipment include:

oxygen cylinder　氧气瓶

oxygen generator　制氧机

oxygen system fittings　氧气系统配件

filler valve　充气阀

check valve　单向阀

shutoff valve　关断阀门

pressure reducer valve　节流阀

pressure relief valve　减压阀

regulator　调节器

pressure gage　压力表

oxygen mask　氧气面罩

In simple form a basic continuous flow oxygen system is illustrated in Figure 4-31.

mask outlet　面罩出口

pressure reducing valve　减压阀

pressure gauge　压力表

line valve　送气管阀门
filter　过滤器
shutoff valve　关断阀门
cylinder　气缸
check valve　单向阀
charging valve　充气阀
charging connection　充气接头

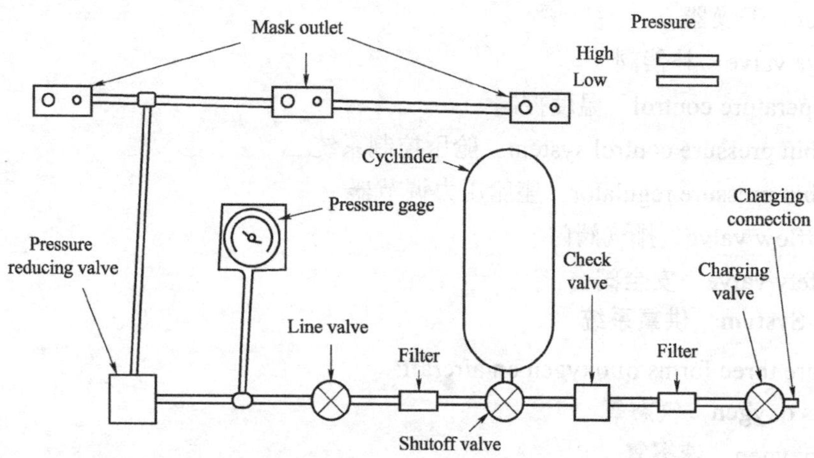

Figure 4-31

6. Ice and Rain Protection　防冰和防雨

anti-icing using combustion heater　燃气加热器防冰
　located in the wing and empennage
anti-icing using exhaust heater　废气加热器防冰
　located in the wing and tail leading edges
anti-icing using engine bleed air　发动机放气防冰
　located in the leading edges, windows and windshields
windshield and carburetor alcohol deicing system　挡风玻璃及汽化器除冰系统
　located in the windshields and carburetor
Pitot tube anti-icing　皮托管防冰
　located in the Pitot tube
water and toilet drain heater　水及盥洗放泄加热器
　located in toilet drain lines, water lines, drain masts and waste water drains
windshield wiper　挡风玻璃雨刷
pneumatic rain removal system　气源除雨系统
　located in windows and the windshield
windshield rain repellent　挡风玻璃排雨剂

7. Fire Protection System　防火系统

(1) fire detection system　火警探测系统

rate-of-rise detector　差温式火灾探测器

radiation sensing detector　热辐射传感探测器

overheat detector　过热探测器

smoke detector　烟雾探测器

carbon monoxide detector　一氧化碳探测器

combustible mixture detector　可燃混合物探测器

fiber-optic detector　光学纤维探测器

(2) Three detector systems in common use are:

thermal switch system　热电门系统

thermocouple system　热电偶系统

continuous-loop detector system　连续环路探测器系统

(3) fire extinguishing system　灭火系统(Figure 4-32)

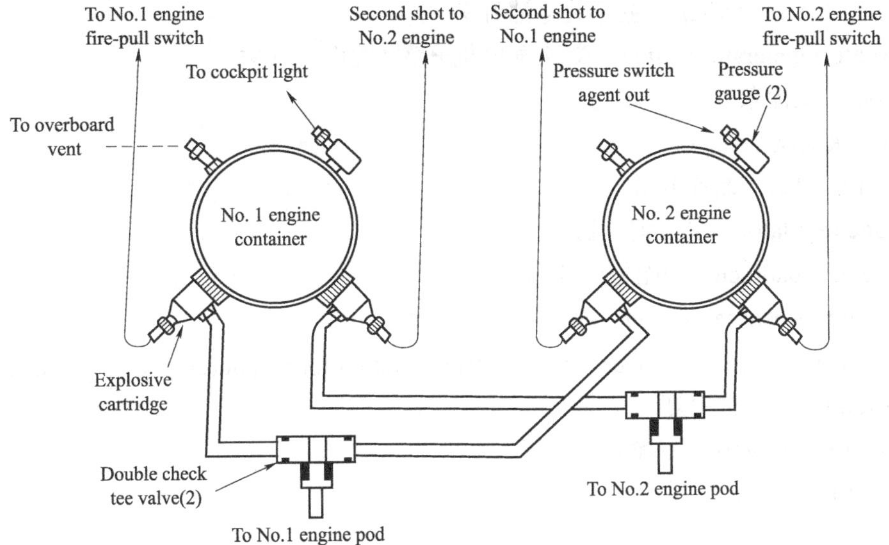

Figure 4-32

fire-pull switch　火警拉响开关

overboard vent　舷外排气孔

explosive cartridge　爆破筒

double check tee valve　双向三通阀

engine pod　飞机引擎基座

engine container　发动机箱

pressure gauge　压力计

pressure switch agent out　压力开关灭火剂喷出

8. Pneumatic System　气源系统

Pneumatic systems operate a great deal like hydraulic systems, except they employ air instead of a liquid for transmitting power. Pneumatic systems are sometimes used for:

brake 刹车
opening and closing doors 舱门开关
driving hydraulic pump 液压泵驱动
alternator 交流发电机
starter 起动机
water injection pump 注水泵
operating emergency device 应急装置操作

9. Electrical System 电气系统

Airplanes are equipped with either a 14- or 28- volt direct-current electrical system. A basic airplane electrical system consists of the following components:

alternator / generator 交流发电机/发电机
battery 电池
master / battery switch 主开关电池开关
alternator/ generator switch 交流发电机开关 / 发电机开关
bus bar 汇流条
fuse 保险丝
circuit breaker 断路开关
voltage regulator 电压调节器
ammeter / loadmeter 电流表/载荷表
associated electrical wiring 相关的连接电线

Equipment that commonly uses the electrical system for its source of energy includes:

position lights 位置灯
anti-collision lights 防撞灯
landing lights 着陆灯
taxi lights 滑行灯
interior cabin lights 舱内指示灯
instrument lights 仪表灯
radio equipment 无线电设备
turn indicator 转弯指示器
fuel gauges 燃油表
electric fuel pump 电动燃油泵
stall warning system 失速警告系统
Pitot heat 皮托管加温
starting motor 发动启动机

10. Power Plant 动力装置

reciprocating engine 活塞发动机(往复式发动机)
turbine engine 涡轮发动机(Figure 4-33)
air inlet or intake section 进风口

compressor 压缩机

combustion chambers / burner section 燃烧室

turbine 涡轮机

exhaust 排气口

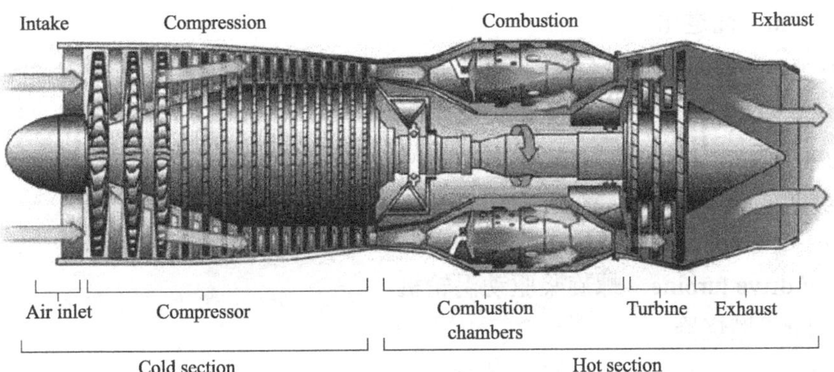

Figure 4-33

Aircraft turbine engines may be classified as turbojet, turbofan, turboprop and turboshaft.

turbojet engine 涡轮喷气发动机(Figure 4-34)

fuel injector 燃料喷射器

nozzle 喷嘴

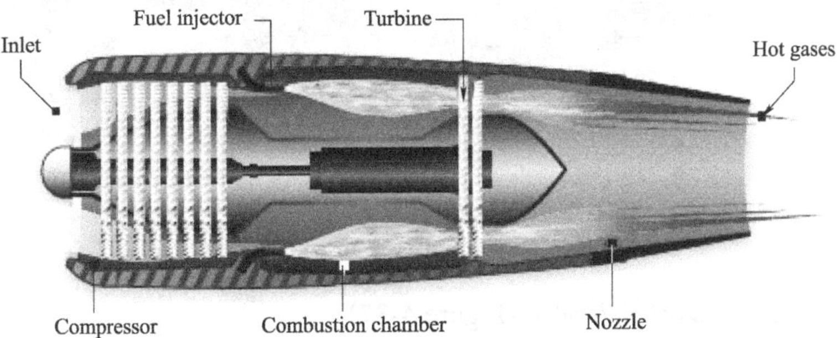

Figure 4-34

turbofan engine 涡轮风扇发动机(Figure 4-35)

fan section 风扇

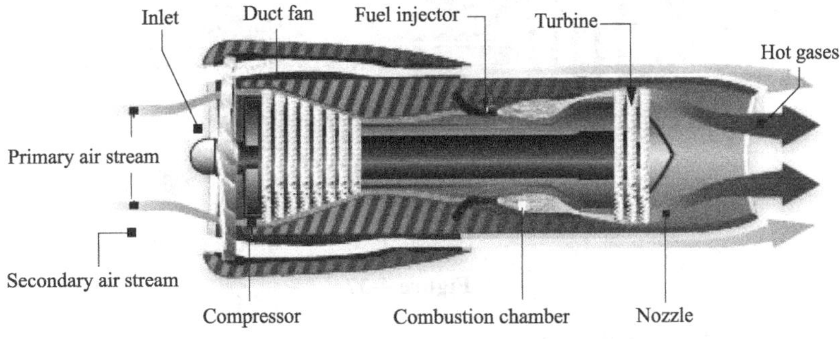

Figure 4-35

compressor section　压缩机
bypass duct　外涵道
inside air duct　内涵道
bypass engine　内外涵式喷气发动机
bypass ratio　涵道比
high bypass　高涵道
turboprop engine　涡轮螺旋桨发动机(Figure 4-36)
gear box　齿轮箱
fuel injector / fuel spray　燃油喷雾
compressor drive turbine　压缩机驱动涡轮机
propeller drive turbine　螺旋桨驱动涡轮机
combustor　燃烧器
reduction gears　减速齿轮，减速装置
propeller　螺旋桨

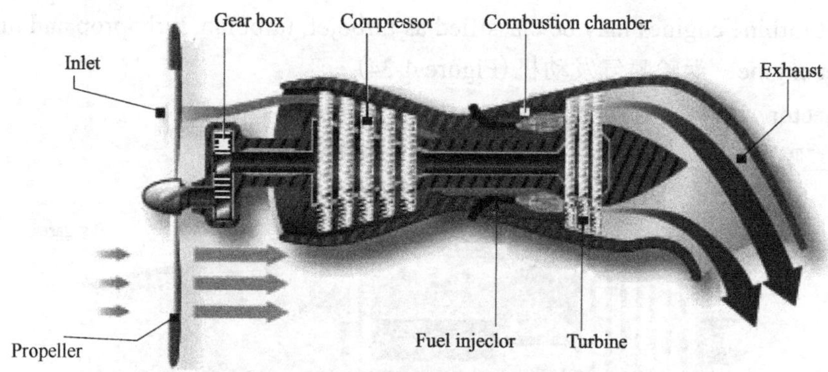

Figure 4-36

turboshaft engine　涡轮轴发动机(Figure 4-37)
power shaft　传动轴

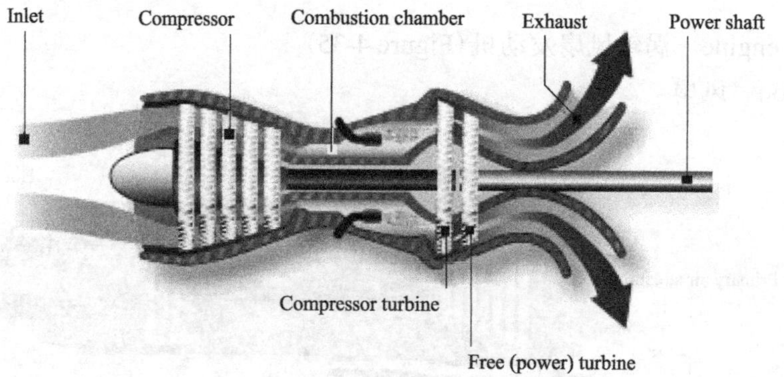

Figure 4-37

ramjet engine　冲压喷气发动机(Figure 4-38)

fuel nozzle 燃料喷嘴

fuel pumps and controls 燃料泵及控制器

igniter-flameholder 点火器火焰稳定器

exhaust 排气口

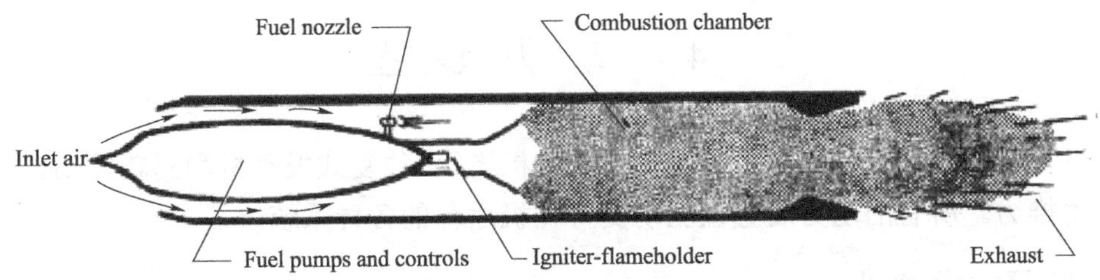

Figure 4-38

第4章 飞机系统

4.1 动力装置

飞机发动机或动力装置产生推力来推动飞机飞行。往复式发动机和涡轮发动机与螺旋桨配合产生推力。所有的动力装置也驱动支持飞机运行的各种系统。

4.1.1 往复式发动机

大多数小型飞机的设计都采用往复式发动机(活塞发动机)。往复式发动机这个名字来源于活塞的前后往复运动。往复运动产生了有效的机械能。

往复式发动机进一步分类如下：

(1)根据气缸排列和曲轴的位置关系，分为星型、直列式、V型和对置式。

(2)根据运行周期，分为二冲程和四冲程发动机

(3)根据冷却方式，分为水冷发动机和风冷发动机。

星型发动机广泛应用于第二次世界大战期间，其中有许多目前还在发挥作用。该发动机是一种单排或多排气缸围绕曲轴环形排列的往复式发动机，其主要优势是具有较好的功重比。

直列式发动机的截面相对较小，但是它们的功重比相对较低。另外，风冷直列式发动机最后面的气缸接收到的冷空气较少，因此该发动机的气缸数量通常为四个或者六个。V型发动机比直列式发动机提供的功率更大，且仍然保留了一个小截面。随着发动机设计的不断创新，人们研发出了水平对置式发动机。该发动机一直是小型飞机上最常用的往复式发动机。这种发动机的气缸数为偶数，因为曲轴箱一侧的气缸和另一侧的气缸一一对应。这些发动机大多数为风冷式，通常安装在飞机固定机翼的水平位置。因为对置式发动机有相对较小的轻型曲轴箱，所以其功重比较高。此外，气缸的紧凑排列缩减了发动机的截面，流线型安装使气动阻力降到最低。

火花点火往复式发动机的主要部分包括气缸、曲轴箱和附件壳。气缸内有进气/排气阀、火花塞和活塞，曲轴和曲轴连杆位于曲轴箱内。磁电机通常位于发动机附件壳内部，如图4-1所示。

四冲程往复式发动机通过气缸的四冲程循环方式将燃油的化学能转化为机械能。这四个独立的活塞冲程为吸气、压缩、做功、排气，如图4-2所示。

(1)吸气冲程从活塞向下运动开始。开始时，进气阀打开，气缸内吸入燃油/空气混合物。

(2)压缩冲程从进气阀关闭，活塞朝气缸顶部移动开始。在循环的这个阶段，可以从点燃的油/气混合物中获得更多能量。

(3)做功冲程从油/气混合气体被点燃开始。该过程会使气缸内压力极大增加，推动活塞离开气缸顶部向下运动，产生使曲轴旋转的动力。

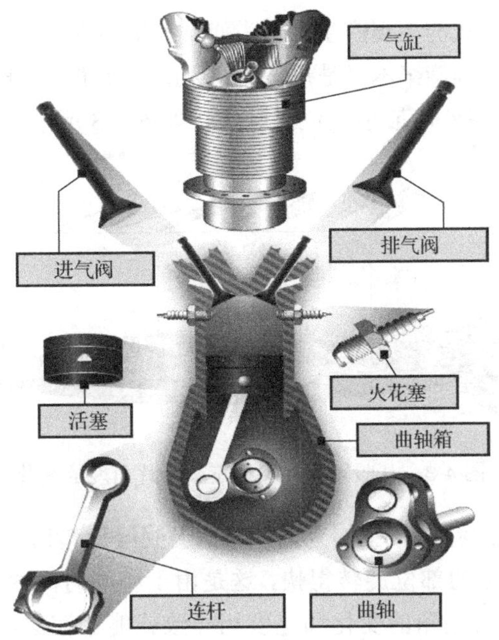

图 4-1　火花点火往复式发动机的主要部分

(4) 排气冲程用于清除气缸中燃烧后的气体。这个冲程发生在排气阀打开，活塞再次开始朝气缸顶部移动的时候。

即使在发动机以相对较低的速度运行时，每分钟四冲程循环也要发生几百次。在四缸发动机中，每个气缸在不同的冲程运行，如图 4-2 所示。曲轴的连续旋转是由每个气缸的做功冲程的精确定时来维持的。发动机的连续运行依赖于辅助系统的同时作用，包含进气、点火、燃油、润滑、制冷和排气系统。

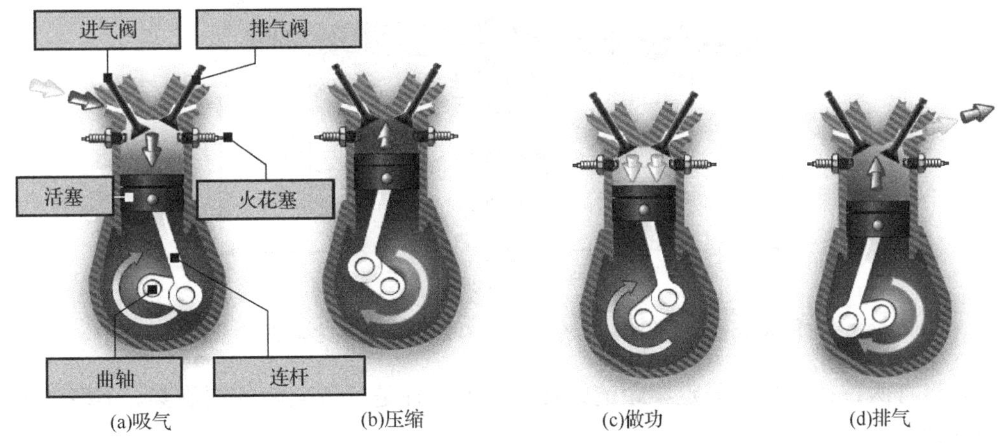

图 4-2　图中箭头指示的四冲程循环中的曲轴和活塞运动的方向

4.1.2　螺旋桨

螺旋桨是一种旋转的翼型，受诱导阻力、失速和适用于任何翼型的其他空气动力学原理的影响。螺旋桨提供必要的推力(有时是拉力)，使飞机在空中飞行。发动机的动力使螺旋桨旋转，螺旋桨产生推力的方式与机翼产生升力类似。产生的推力大小取决于桨叶的形态、螺旋桨叶迎角和发动机的转速。螺旋桨本身是扭转的，因此桨叶角从毂轴到叶尖是变化的。最大桨叶迎角或者最大桨距在毂轴处，最小值位于叶尖，如图 4-3 所示。

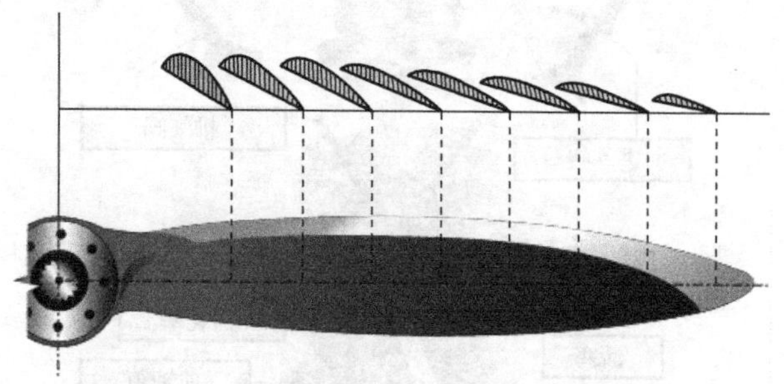

图 4-3　从毂轴到叶尖螺旋桨叶片角度的变化

扭转是为了使从毂轴到叶尖产生的升力一致。当桨叶旋转时，桨叶的不同部分实际速度不同。桨叶尖部比靠近毂轴的部位旋转得快，这是由于相同时间内叶尖要旋转的距离比毂轴附近的长。从毂轴到叶尖安装角的变化和速度相一致就能够在桨叶长度上产生一致的升力。如果螺旋桨叶设计成与整个长度上的安装角相同，那么螺旋桨效率将会降低。因为随着空速的增加，靠近毂轴附近的部分会有负迎角，叶尖会发生失速，如图 4-4 所示。

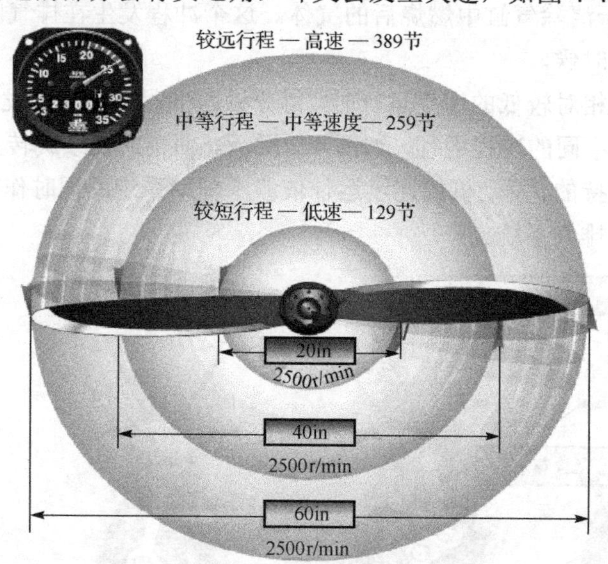

图 4-4　螺旋桨叶片各个部分运行距离和速度之间的关系

螺旋桨有两种：一种是固定桨距的螺旋桨，另一种是可调桨距的螺旋桨。小型飞机装配其中的一种。

1．定距螺旋桨

具有固定叶片角度的螺旋桨称为定距螺旋桨，这种螺旋桨的桨距是制造商设定的，不能改变。只有在一定的空速和转速组合下这种螺旋桨才能获得最高的效率。这种固定桨距对于巡航或者爬升都不是最理想的。因此，飞行器的各类性能会受到一定的限制。固定桨距螺旋桨只适用于飞行器质量轻、结构简单、成本低的情况。

定距螺旋桨有两种类型，即爬升螺旋桨和巡航螺旋桨。飞机安装爬升螺旋桨还是巡航螺旋桨，取决于其预期用途。爬升螺旋桨的桨距较小，因此阻力更小，转速更高，马力更大，这增强了起飞和爬升的性能，但是在巡航飞行时性能会有所减弱。巡航螺旋桨的桨距较大，因此它的阻力更大，从而导致转速较慢和功率能力较低，降低了起飞和爬升性能，但是提高了巡航飞行效率。

螺旋桨通常安装在轴上，这个轴可能是发动机曲轴的延伸。这时，螺旋桨的转速和曲轴的转速相同。对于某些发动机，螺旋桨安装在其曲轴经齿轮传动的轴上。这时，曲轴的转速和螺旋桨的转速不同。

对于定距螺旋桨，转速表是发动机功率的指示器。

如图 4-5 所示，转速表的刻度以 100r/min 为单位，直接显示发动机和螺旋桨的转速。转速表有颜色标记，其中绿色弧线表示最大连续运行转速。一些转速表还标记了对发动机或螺旋桨的限制条件。对转速表的标记有不清楚的地方可参考其制造说明书。

图 4-5 转速表显示发动机转速

发动机的转速由油门控制，油门的开关可调节进入发动机的燃油或燃气的量。在一定的飞行高度，转速表读数越大，发动机输出功率越大。

当飞机飞行高度增加时，转速表显示的可能不是发动机正常的输出功率。例如，2300r/min在5000英尺高度时产生的马力比在海平面时产生的马力要小。这是因为发动机的功率输出和空气密度有关。空气密度随高度增加而降低，而空气密度的降低（即高密度高度）导致了发动机输出功率的降低。当高度变化时，必须要改变油门的位置才能维持相同的转速。也就是说，当高度增加时，必须加大油门，以维持和低高度时相同的转速。

2. 变距螺旋桨

变距螺旋桨是恒速螺旋桨的前身。这种螺旋桨只能在发动机停转时在地面调节，而不能在飞行中调节，也被称为地面可调螺旋桨。到20世纪30年代，航空先驱发明家们奠定了自动调节螺距机制的基础，因此变距螺旋桨也指可在飞行中调节桨距的现代恒速螺旋桨。

最初的变距螺旋桨系统只能提供两种桨距设定，即低桨距和高桨距。现在大多数变距螺旋桨都能在一个范围内调节桨距。

恒速螺旋桨是最常见的变距螺旋桨。它的桨距在飞行中可以通过调节器自动改变，进而可以忽略空气载荷的变化而使螺旋桨维持恒定转速。恒速螺旋桨的主要优点在于它能在较大空速和转速的组合范围内将大部分制动马力转换为推进马力。恒速螺旋桨能在特定条件下选择相对应效率最高的发动机转速，因此它比其他螺旋桨效率更高。

装配恒速螺旋桨的飞机有两项控制，即油门控制和螺旋桨控制。油门控制功率输出，螺旋桨调节发动机转速，这两项控制共同调节转速表上显示的螺旋桨转速。

一旦选择了一个特定的转速，调节器会自动调节相应的螺旋桨桨叶角以保持所选的转速。例如，巡航飞行期间设定了需要的转速之后，空速的增加或者螺旋桨载荷的降低会导致螺旋桨为维持所选的转速而增大桨叶角。相应地，空速降低或者螺旋桨载荷增加会导致螺旋桨桨叶角减小。

螺旋桨的恒速范围就是恒速螺旋桨的桨叶角范围，由其高低桨距止位确定。只要螺旋桨桨叶角位于恒速范围内，且不超出任何一个桨距止位，发动机转速就能维持恒定。如果螺旋桨桨叶到达桨距止位，发动机转速就会随空速和螺旋桨载荷的变化而相应增加或者降低。比如一旦确定了一个特定的转速，如果飞机速度降低到足够使螺旋桨桨叶旋转直到低桨距止位，那么只要空速再次降低就会导致发动机转速降低，就像安装了固定桨距螺旋桨一样。当装有恒速螺旋桨的飞机加速到较快的速度时还会发生同样的情况。随着飞机加速，螺旋桨桨叶角增加，以维持选定的转速直到高桨距止位。一旦达到止位，桨叶角将不再增加，发动机转速降低。

图 4-6 进气压力表显示发动机功率输出

在装配恒速螺旋桨的飞机上，功率输出由油门控制，并由图4-6所示的进气压力表显示。这个仪表测量进气歧管中油气混合物的绝对压力，更准确地说是测量歧管绝对压力(MAP)。在恒定转速和高度下，油门产生功率的大小与流到燃烧室的油气混合物有直接关系。当加大油门时，流到发动机的油气混合物就越多，因此，歧管绝对压力增加。当发动机不运行时，进气压力表显示周围空气压力(如29.92in Hg)。当发动机起动后，歧管压力显示值将会降低到一个低于周围空气压力的值(如12in Hg)。相应地，当发动机出现故障或者功率损耗时，相比于发生故障时飞机所处高度的空气压力，进气压力表指示数增大。

进气压力表用色标来指示发动机的可运行范围。进气压力表盘上有一段绿色弧线表示正常运行范围，红色径向线指示歧管压力的上限值。

对于任何给定的转速，都有一个不能超过的歧管压力值。如果对应转速下的歧管压力过大，气缸内部的压力就会过大，导致气缸受到过大的应力。如果频繁地出现这种情况，那么这个应力将会使气缸组件变松，最终导致发动机故障。因此，一般规定歧管压力应小于转速。

飞行员可以通过不断关注转速来避免气缸过应力的状况，特别是在增加歧管压力时。飞行员应当遵循发动机说明书的指导设定转速，这样歧管压力和转速之间才能够维持合适的关系。

4.2 进 气 系 统

进气系统将外部空气吸入并与燃油混合，混合物被送到气缸内进行燃烧。外部空气从发动机整流罩前部的进气口进入进气系统。这个进气口通常包含一个阻止灰尘和其他外部物体进入的空气过滤器。由于过滤器偶尔会堵塞，因此设置有备用空气源。通常情况下，备用空气来自发动机整流罩内部，并且绕过阻塞的空气过滤器。一些备用空气源采用自动操作，而另一些则需要手动操作。

小型飞机通常使用两种类型的进气系统：

(1)汽化器系统。燃油和空气在进入进气歧管之前，先在汽化器中混合。

(2)燃油喷射系统。燃油和空气在进入每个气缸之前迅速混合，或直接将燃料注入每个气缸。

4.3 燃 油 系 统

燃油系统用来持续地为发动机提供来自油箱的洁净燃油，其作用是保证在各种飞行状态和条件下(如发动机功率、高度、飞行姿态等)都能为发动机提供所需的燃油量。小型飞机上使用了两类常规的燃油系统，即重力馈送系统和燃油泵系统。

重力馈送系统利用重力把燃油从油箱输送到发动机。例如，在上单翼飞机中，油箱安装在机翼里。在这种情况下，油箱置于汽化器之上，燃油由于重力经过系统送到汽化器，如图 4-7 所示。如果飞机的设计导致不能用重力输送燃油，就需要安装燃油泵，例如，在下单翼飞机的机翼中，油箱处于汽化器下方，如图 4-8 所示。

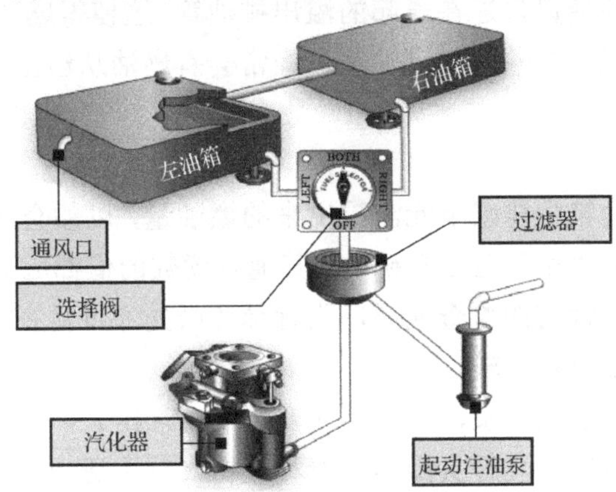

图 4-7 重力馈送系统

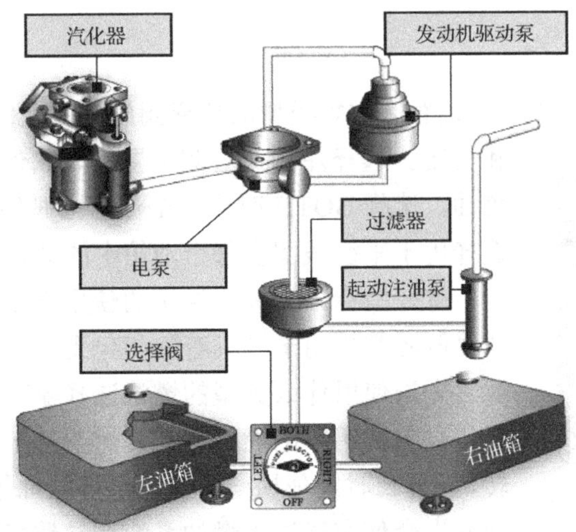

图 4-8 燃油泵系统

装有燃油泵系统的飞机使用两组油泵。主泵系统由发动机驱动，并配有电动辅助泵，辅助泵用于发动机起动和发动机泵发生故障时，也称为增压泵，为燃油系统增强可靠性。电驱动辅助泵由驾驶舱中的开关控制，如图 4-8 所示。

4.3.1 起动注油泵

重力馈送和燃油泵系统都有可能装配起动注油泵。在起动发动机之前，起动注油泵将油箱中的燃油直接汽化送入气缸。天气较冷时，如果没有起动注油泵，则没有足够的热量来汽化汽化器中的燃油，发动机很难起动。起动注油泵在不使用时要固定好其位置。如果旋钮可以自由活动，那么在飞行中它会被振出来，导致注油过量。为避免注油过多，请阅读飞机上的注油说明。

4.3.2 油箱

油箱通常位于飞机的机翼内，可以通过机翼顶部的加油口加注，其由一个盖子盖住。油箱设计有通风口以平衡油箱内外气压。有的飞机在油箱盖或机翼表面伸出有通气口。油箱包括一个单独的或者和油箱通风口连在一起的溢出排油管。这使得燃油在温度升高而膨胀时不会损坏油箱本身。如果天气较热时加满油箱，经常会有燃油从溢出排油管流出。

4.3.3 油量计

油量计显示每一个油箱中传感单元测量出来的燃油量，以加仑或者磅为单位。飞机认证规则表示油量计在显示为“空”时是精确的，其他有读数的值都应该经过校验。因此不能只依赖油量计的示数。在飞行前的检查期间，飞行员务必先通过目视检查每一个油箱的油量，然后跟对应的油量表读数进行比较。

如果燃油系统中安装了燃油泵，则会相应安装一个油压表，这个表显示油管的压力值。飞机飞行时的油管压力可以在飞行员操作手册和飞机飞行手册中查找，或者根据仪表刻度盘上的色标确定。

4.3.4 燃油选择器

燃油选择器用于选择不同的油箱。常规类型的选择阀门有四个位置：左侧、右侧、两侧和关闭。选择左侧或右侧位置代表只使用左边或者右边油箱的燃油，选择 BOTH 时由两个油箱同时供油。左侧和右侧位置的选择用于平衡残留在每个油箱中的油量，如图 4-9 所示。

图 4-9　燃油选择器

有一些标语会提醒油箱的使用限制，如“仅限水平飞行”或“两者都”用于飞机着陆和起飞。

无论使用什么类型的燃油选择器，都应该密切关注燃油消耗，以免油箱中的燃油用光。燃油的耗尽不仅会导致发动机停转，而且长时间使用一个油箱会导致油箱之间的燃油载荷失衡。同时也会使空气进入燃油系统，导致气阻。一旦出现气阻的情况，则很难再起动发动机。在燃油喷射型发动机上，燃油温度较高，导致燃油在进入气缸之前在油管中汽化。

4.3.5　燃油过滤器、油底壳、排油管

燃油在离开油箱，进入汽化器之前会经过一个过滤器。这个过滤器用于清除灰尘和系统中可能存在的其他沉积物。由于这些污染物比航空燃油重，因此会沉淀到过滤器部件底部的油底壳中。油底壳是燃油系统或者油箱中的低位置点。燃油系统可配备有油底壳、燃油过滤器和油箱排油管。

飞机每次飞行前都应将燃油过滤器中的油排干，从过滤器取出燃油样本，并目视检查其中的水和污染物。

沉积器中有水时较危险，因为在天气较冷时水会结冰从而堵塞油管。天气较热时，水会流进汽化器，导致发动机停止工作。如果油底壳中有水，则意味着油箱中残余的水份更多，需要将其完全排出。在发动机燃油系统中的水和污染物被排尽之前，飞机禁止起飞。

4.3.6　燃油等级

航空汽油的等级用辛烷值或品度值标定，辛烷值也被视为发动机气缸中燃油混合物的抗爆指数或抗爆震性能。汽油的等级越高，燃油能承受的不产生爆燃情况下的压力就越大。由于较低等级的燃油可在低温下点燃，因此可以用在低压缩比的发动机上。而较高等级的燃油在较高温度下点燃且不能过早点燃，因此用在高压缩比的发动机上。如果没有匹配等级的燃油可用，那么可使用高一等级的燃油替代。禁止使用低一级的燃油，因为这会导致气缸盖温度(也就是机油温度)超出发动机正常运行时的温度范围，从而引发爆燃。

目前有几种等级的航空汽油可用。工作人员必须仔细确保特定类型的发动机使用正确的航空汽油等级。飞机飞行手册和飞行员操作手册、驾驶舱内的标牌以及加油口盖旁都有相对应的等级标注。汽车使用的汽油禁止用于飞机发动机，除非飞机的发动机已经按照美国联邦航空管理局(FAA)颁发的补充型号合格证(STC)改装过。

当前识别用于往复式发动机的飞机航空汽油的方法是根据汽油的辛烷值和品度值。飞机使用的汽油等级分为 AVGAS80、100 和 100LL。尽管 AVGAS 100LL 的性能和 100 一样，但“LL”表示它的含铅量低。涡轮发动机飞机的燃油是用 JET A、JET A-1 和 JET B 分类的。喷气机使用的燃油主要是煤油，且有特殊的煤油气味。给飞机使用正确的燃油非常重要，因此通过在燃料中添加染料来帮助识别燃油的类型和等级，如图 4-10 所示。

除了给燃油标记颜色之外，机场的燃油设备也贴有不同的色标。所有航空汽油的标记是写在红色背景上的白字，而涡轮机燃油的标记是写在黑色背景上的白字。

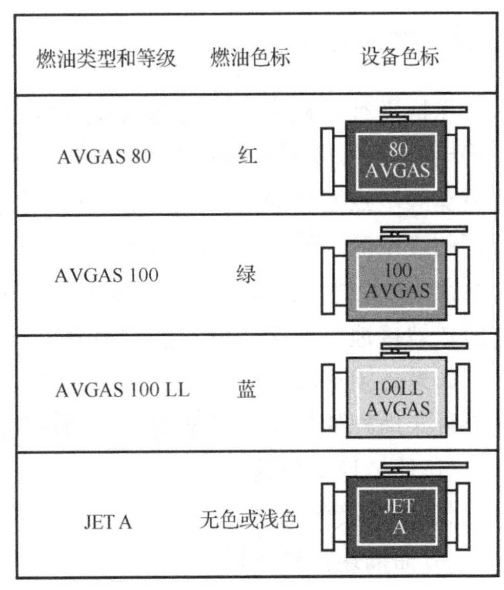

燃油类型和等级	燃油色标	设备色标
AVGAS 80	红	80 AVGAS
AVGAS 100	绿	100 AVGAS
AVGAS 100 LL	蓝	100LL AVGAS
JET A	无色或浅色	JET A

图 4-10　航空燃料的颜色编码系统

4.3.7　燃油污染

由于燃油污染导致动力装置而引发的事故大多数归因于：

(1) 飞行员在飞行前没有进行充分的检查。

(2) 使用来自小油箱和油桶中过滤不当的燃油。

(3) 存放飞机时油箱未加满。

(4) 燃油系统缺乏正确的保养。

首先将油箱油底壳中的燃油通过燃油过滤器排放口排出并放入一个透明容器内，然后检查燃油中的水和污染物。当燃油过滤器使用时，只有从连到油箱的管子排出所有的油后才能看到油箱中的水。这表明水保留在油箱里，没有使燃油从通向燃油过滤器的管道中流出。因此要从燃油过滤器排出足够的油以确保燃油从油箱放出。油量取决于从油箱到排油管之间的燃油管道长度。如果在第一份取样中发现水或者其他污染物，则需要一直取样，直到没有污染物存留。

尽管燃油过滤器的排油已停止，表明没有水的存在，但水箱中仍然有水残留的可能。而残留的水只能通过油箱油底壳放油口排出。

水是主要的燃油污染物。当油箱中的水沉到底部时，燃油中悬浮的水滴可以通过燃油的云状外形或者水与有色燃油的区别来分辨。出于安全考虑，在飞机每次飞行前的检查中，需要对油箱底壳进行排油处理。

飞机每次飞行或当天最后一次飞行后，都需要将油箱加满燃油，以防止油箱中出现水汽凝结。同样为防止水汽凝结，禁止从油罐或油桶中加油。

在偏远地区或紧急情况下，如果没有能足够抗污染的备用加油源，使用麂皮或漏斗过滤燃油的处理方式可能是唯一的解决办法。但是使用麂皮和漏斗也有风险，因为不能保证每次过滤时燃油不被污染。破旧的麂皮则不能过滤水分；即使是一个新的、干净的湿润麂皮也不能过滤水分。大多数仿制麂皮也不能过滤掉水分。

4.3.8　加油程序

飞机飞行过程中空气通过飞机表面产生摩擦和加油过程中燃油流经软管和喷嘴都会产生静电。尼龙、涤纶或者羊毛材质的衣服都易于聚积和释放静电，从人体流向漏斗或喷嘴。为避免静电点燃燃油油雾，打开油箱盖之前接一根地线连到飞机上。因为飞机和加燃料器有不同的静电荷，所以结合这两部分至关重要。通过结合这两部分，静电电荷差得到平衡。在加油前，应先将加油喷嘴与飞机搭接，并在整个加油过程中都要保持搭接。当使用加油车时，它应该被接地至连接飞机的加油喷嘴。

如果必须从油罐或者油桶加油，那么正确的连接方式和接地连接变得重要。油桶应该放在靠近接地杆的位置，且要遵守下列连接顺序：

(1) 油桶连接到地。

(2) 地连接到飞机。

(3) 油桶连接到飞机。

(4) 拿掉油箱盖之前将喷嘴与飞机搭接。

当断开连接时，顺序相反。

燃油流经麂皮增加了静点荷和打火花的危险。飞机必须正确接地，喷嘴、麂皮和漏斗要正确搭接到飞机上。如果使用了油罐，则应该将其与接地板或漏斗连接。任何情况下这个操作都不能使用塑料漏斗或者类似的绝缘容器。

4.4 起 动 系 统

大多数小型飞机使用的是一个直接摇转起动机系统。这个系统包括电源、导线、开关，以及操作起动器的螺线管和起动电动机。大多数飞机的起动机工作时可以自动接通和断开，但是一些老式飞机的起动机是通过飞行员操纵的控制杆来控制的。起动机接通飞机的飞轮，以一个可以起动发动机和保持发动机运行的速度使发动机运转。

起动所需的电力通常由机载电池提供，同时也可以由外部电源提供。当电池开关打开时，电力通过电池螺线管供应到主电力汇流条。起动机和起动机开关都从汇流条获得电流，只有当起动机开关被旋到开始位置，激励了起动螺线管，才能操作起动机。当起动机开关离开开始位置后，起动发动机失去来自螺线管的电力。起动电动机受到保护，以防被发动机通过起动机传动的离合器驱动，离合器能使发动机转得比起动电动机更快。起动电路如图4-11所示。

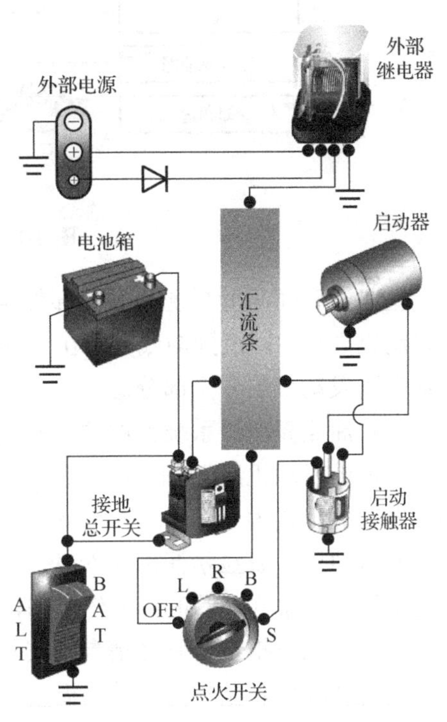

图 4-11 典型的起动电路

起动发动机时，必须严格遵守安全和文明规则。其中最重要的一条是确定没有人靠近螺旋桨。另外，轮子应该使用制动垫块并合上手闸，以避免意外移动带来的危险。为避免损坏螺旋桨和财物，飞机应该停在螺旋桨不能扬起沙粒和尘土的地方。

4.5 润 滑 系 统

发动机润滑系统实现几个重要的功能，包括：

(1) 发动机活动部件的润滑。

(2) 通过减少摩擦来冷却发动机。

(3) 带走气缸的热量。

(4) 提供气缸壁和活塞之间的密封。

(5) 带走污染物。

往复式发动机使用湿式油底壳或干式油底壳润滑系统。在湿式油底壳润滑系统中，润滑油位于油底壳，是发动机整体的一部分。在干式油底壳润滑系统中，润滑油存储在一个独立的油箱里，通过油泵在发动机中循环。湿式油底壳润滑系统如图4-12所示。

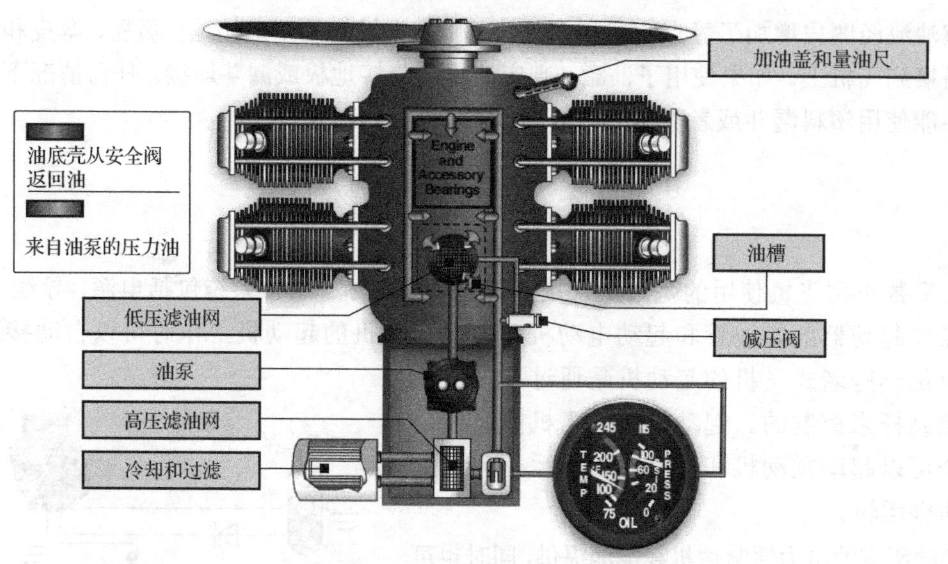

图 4-12 湿式油底壳润滑系统

湿式油底壳润滑系统的主要部件是油泵，它从油底壳中抽油并导流到发动机。润滑油流经发动机之后，回流到油底壳。在一些发动机内，旋转的曲轴提供额外的润滑作用，它将润滑油带到发动机的各个部分。

干式油底壳润滑系统也由油泵提供油压，但是润滑油来源于发动机外部独立的油箱内。润滑油流经发动机后，被回油泵从发动机的不同部分抽回到油箱内。干式油底壳润滑系统能够为发动机提供大量润滑油，因此更适用于大型的往复式发动机。

润滑油压力表显示润滑系统的工作情况。它以磅/平方英寸为单位显示供应到发动机的润滑油压力。绿色表示正常工作范围，红色表示最小压力和最大压力。发动机起动时润滑油压力表上有油压指示。具体请参考飞机飞行手册或者飞行员操作手册。

润滑油温度表显示润滑油的温度。绿色区域表示正常工作范围，红色线表示最大允许温度。和润滑油压力不一样，润滑油温度的变化更为缓慢。在起动一台较凉的发动机时特别明显，可能需要几分钟或者更长时间才能看到温度表显示温度增加。

飞行时需要定期检查润滑油温度，特别是在周围的空气温度较低或者较高时。高温度读数可能表示油管堵塞，润滑油油量变低，润滑油制冷器阻塞或者温度表故障。低温读数可能表示在冷天气运行时润滑油的黏度不合适。

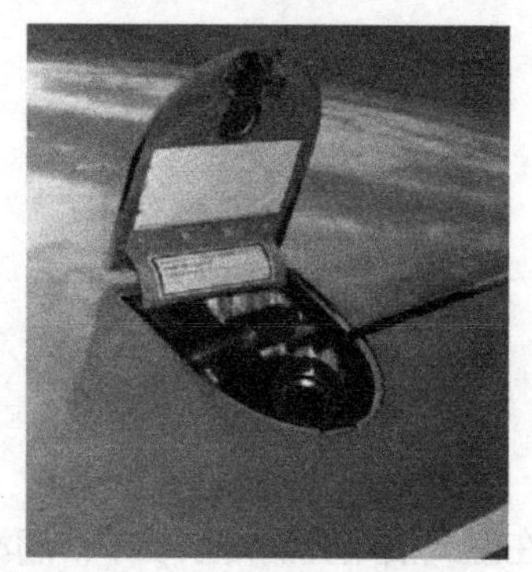

图 4-13 在飞行前经常检查发动机机油油位

润滑油加油盖和量油计(测量润滑油的油量)通常位于发动机整流罩内的操纵板上。如果润滑油量没达到制造商建议的运行油量要求，那么需要增加润滑油。飞机飞行手册和飞行员操作手册或者靠近操纵板边上的标牌会提供正确的润滑油类型和重量信息，包括最小油量和最大油量。如图 4-13 所示。

4.6　发动机冷却系统

气缸内燃烧的燃油产生大量的热量，大部分热量通过排气系统排放出去。同时必须将大部分剩余的热量耗散出去，以防发动机过热。否则，过高的发动机温度会引起功率损耗，过量的燃油消耗，爆燃和发动机严重损坏的情况。

虽然润滑油系统对于发动机的内部冷却很重要，但是发动机外表面的制冷方法也是必需的。尽管一些小型飞机的制冷方式是液冷的，但是大多数是风冷的。

空气通过发动机整流罩前面的开口流向发动机隔舱来实现风冷。导流片引导空气从发动机气缸外的散热片和其他部件上流过，这时空气吸收了发动机的热量。随后热空气通过发动机整流罩的后下方部分的一个或多个开口排出，如图4-14所示。

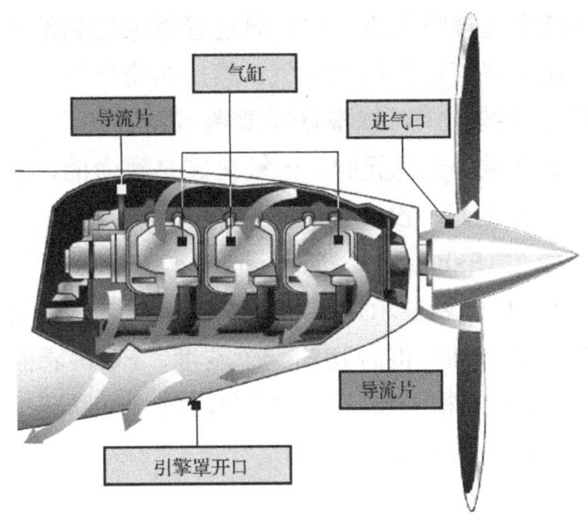

图4-14　发动机冷却时外部空气的帮助

外部空气通过螺旋桨轮毂后面的入口进入发动机的隔舱。导流片将它们导流到发动机最热的部分，主要是装有散热片的气缸，散热片还能增加气缸暴露在气流中的面积。

空气制冷系统在地面运行、飞机起飞，复飞以及其他高功率低空速运行阶段时工作效率不高。相反，高速下降会产生过量空气，使发动机快速冷却，产生这一现象的原因是受温度突然波动的影响。

发动机运行时超过其额定温度会导致功率损耗、润滑油消耗过多以及爆燃，还会导致部件严重损坏，如擦伤气缸壁、损坏活塞和活塞环、烧毁阀门或使其变形。监视驾驶舱中发动机温度仪表能够有效避免高温运行。

在正常运行条件下，未装配通风片的飞机的发动机温度可以通过改变空速或者发动机输出功率来控制。若发动机温度较高，可以通过增加空速或减小功率来降低。

润滑油温度表对发动机温度升高的显示是间接延迟的，但是如果只有这个方法可用的话，也可以用于确定发动机温度。

很多飞机装配了气缸头温度表，这个仪表可以直接显示气缸的温度变化。其刻度以摄氏度或者华氏度为单位，通常带有色标。仪表上的绿色弧线表示正常运行范围，红色线表示最大允许气缸头温度。

为避免气缸头温度过高，可以增加空速、控制注油或降低功率。用这三者中的任意一个办法都可以降低发动机温度。装配了通风片的飞机一般使用通风片来控制温度。通风片是铰接在开口上的盖子，热空气通过它得以排出。如果发动机温度低，可以关闭通风片，从而限制热气流的排出，使发动机温度升高。如果发动机温度高，可以打开通风片，让更多的气流通过冷却系统，从而降低发动机温度。

4.7 排 气 系 统

发动机排气系统把燃烧完的气体排出机外，为机舱提供热量，为挡风玻璃除霜。排气系统由连接在气缸上的排气导管、消声器和消声器护罩组成。废气通过排气阀门排出气缸，然后经过排气管系统排放到大气中。

为了给机舱供热，外部空气被吸入进气口，通过管道输送到消声器周围的护罩。消声器由现存的废气加热，进而加热消声器周围的空气。然后热空气被管道输送到机舱，用于供热和除霜。供热和除霜由驾驶舱控制，可以根据需要调节。

废气中含有大量的一氧化碳，无色无味。一氧化碳是致命的，但实际上很难察觉到它的存在。因此排气系统必须运行良好，没有裂缝。

一些排气系统有排气温度(Exhaust Gas Temperature，EGT)探头。这个探头将排气温度传送到驾驶舱的仪表上。EGT 仪表显示排气歧管中的废气温度。这个温度随进入气缸的油气混合比而变化，可以作为调节油气混合物的一个基准。EGT 表能够非常精准地指示正确的油气混合设定。当使用 EGT 来精确设定油气混合比时，可以减少燃油消耗。关于具体流程，可以参考制造商对油气混合的建议。

4.8 电 气 系 统

大多数飞机都配备了 14V 或者 28V 直流电气系统。一个基本的飞机电气系统包含下列组成部分：

(1) 交流发电机/发电机；

(2) 电池；

(3) 主开关/电池开关；

(4) 交流发电机开关/发电机开关；

(5) 汇流条、熔断器和断路器；

(6) 调压器；

(7) 电流表/载荷表；

(8) 相关电线。

发动机驱动的交流发电机或者发电机为电气系统提供电流，也为电池维持足够电荷量。存储在电池中的电能为起动发动机提供电源，也可以在交流发电机或发电机失效的时候提供有限电力。

大多数直流发电机在低转速时不能提供足够的电流来运行整个电气系统。因此，发动机低转速运行期间，电力需求必须通过电池来满足，但电池消耗较快。

交流发电机与发电机相比有几个优势。交流发电机通过交流电能够产生足够的电流来运作整个电气系统，甚至在发动机转速较低时，产生的交流电流可以转换成直流。交流发电机的电力输出在发动机的较大范围转速内更加恒定。

一些飞机有连接地面电源设备(GPU)的插座，可为起动发电机提供电能。特别是在冷天气起动发电机时，这些插座非常有用。可以根据制造商对地面电源设备(GPU)使用的建议来起动发动机。

电气系统用主开关打开或关闭。当主开关旋到打开(ON)位置时，除点火系统外，其他所有电气设备电路都会获得电能。通常使用电气系统作为其能源来源的设备包括：

(1)位置光信号；

(2)防撞灯；

(3)着陆灯；

(4)滑行灯；

(5)机舱内灯；

(6)仪表灯；

(7)无线电设备；

(8)转弯指示器；

(9)燃油表；

(10)电动燃油泵；

(11)失速警告系统；

(12)皮托管加热；

(13)起动电动机。

很多飞机配备了电池开关，用于控制飞机的电力，方式类似于主开关。另外，安装了交流发电机开关，这个开关可以让飞行员在交流发电机发生故障时，将交流发电机排除在电气系统之外(图4-15)。

当交流发电机开关位于关闭(OFF)位置时，全部电力负荷都在电池上。因此，要关闭所有不必要的电子设备以保存电池电量。汇流条作为飞机电气系统的接线端子，用于连接主电气系统和使用电力作为动力源的设备。这简化了布线系统并且提供分布于系统的常用电压接入点。电气系统中使用的熔断器或者断路器用于保护电路和设备，以防电力过载。

图4-15 在这个主开关，左半边是交流发电机，右半边是电池

飞机上应该备有适当安培极限的备用保险丝来替换失效或损坏的保险丝。断路器具有和保险丝相同的功能，如果电气系统发生过载情况，断路器可以手动复位，而不是替换。保险丝或者断路器面板上的标牌以名字显示电路，同时显示电流极限值。

电流表用于监控飞机电气系统的性能。当交流发电机/发电机提供足够的电力供应时，电流表会有所显示，也会指示电池是否正在充电。

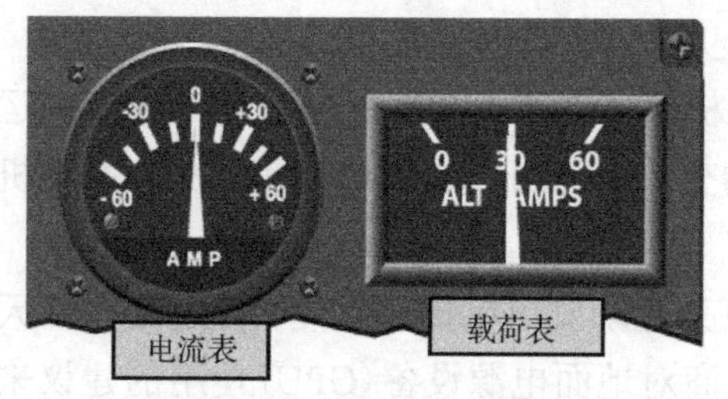

图 4-16　电流表和载荷表

电流表设计以表盘中央为零点，其左右两边为正负指示，如图 4-16 所示。当左侧电流表(图 4-16 左侧的仪表)指针在正偏转的一边(指针右偏)时，它表示电池的充电率。读数为负意味着从电池汲取的电流多于被替换的电流。全程负偏转表示交流发电机/发电机有故障。全程正偏转表示调压器有故障。无论哪种情况，都需要参考飞机飞行手册或者飞行员操作手册，采取正确的措施。

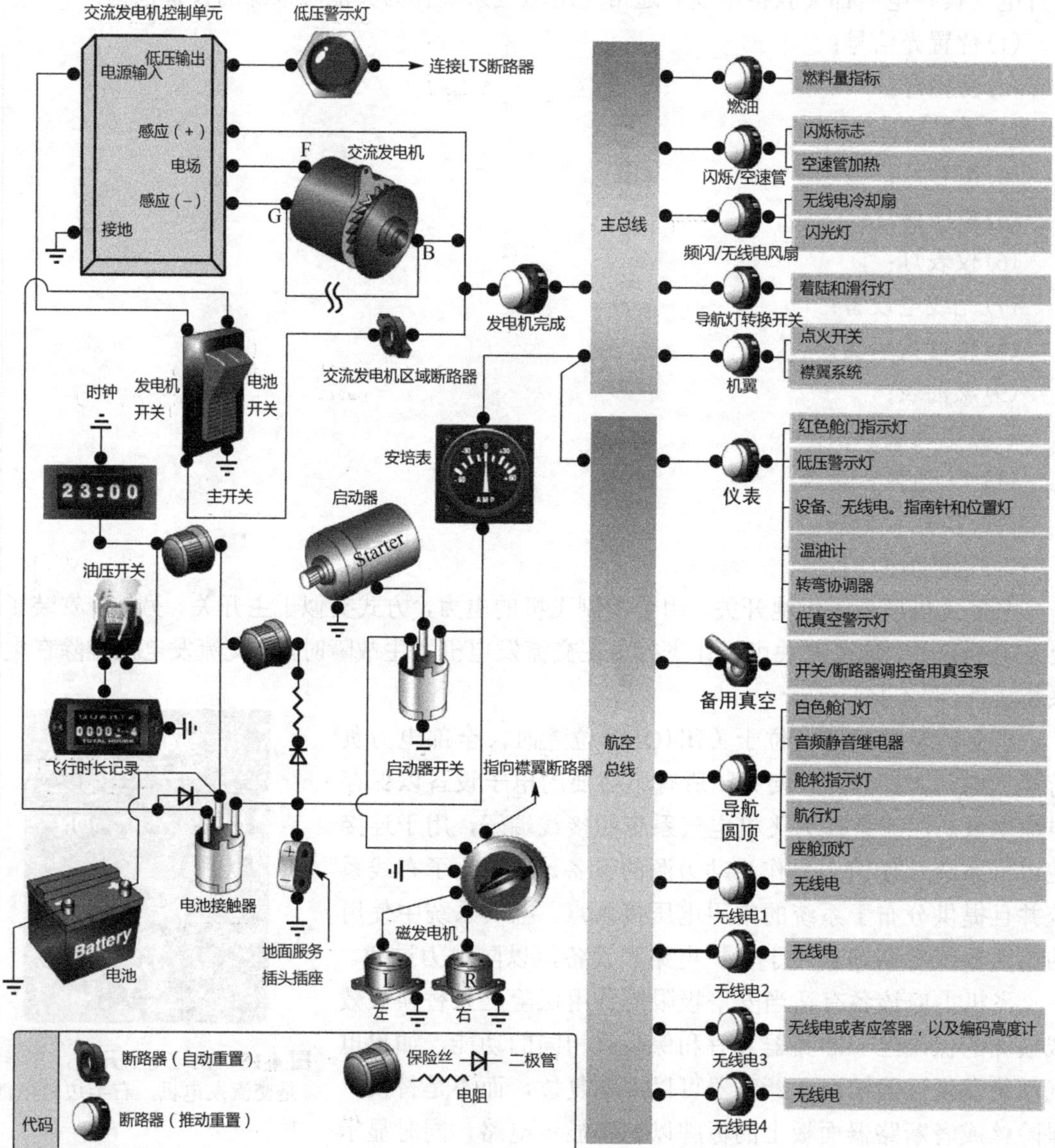

图 4-17　电气系统目录

不是所有的飞机都装配了电流表。一些飞机有警告灯，当灯亮时，它表示发电机/交流发电机存在故障而导致电气系统放电。可以参考飞机飞行手册或者飞行员操作手册(图 4-17)，采取正确的措施。

另一个电力监视指示器是载荷表。这种仪表有一个从 0 开始的量程，用于显示交流发电机/发电机上的载荷。载荷表通过电子配件和电池来反映载荷基于电气系统发电量的总百分比。当所有电力组件关闭时，它只反映电池需要的充电电流的大小。

电压调节器通过稳定发电机或者交流发电机的电力输出来控制电池的充电速度。发电机/交流发电机的输出电压应该比电池电压高。例如，12V 的电池应该用大约 14V 的发电机/交流发电机系统充电。电压差使电池不断充电。电流表和载荷表如图 4-16 所示。

4.9　液 压 系 统

飞机上使用的液压装置有多种用途，这取决于飞机的复杂性。例如，在小型飞机上，液压系统经常用于操作车轮制动器、可伸缩起落架和一些恒速螺旋桨。在大型飞机上，液压装置用于飞行控制面、襟翼、扰流板和其他系统的控制。

基本的液压系统由油箱、泵(手动、电力或者发动机驱动)、保持液体清洁的过滤器、控制流向的选择阀，缓解超压的安全阀和作动筒组成(图 4-18)。

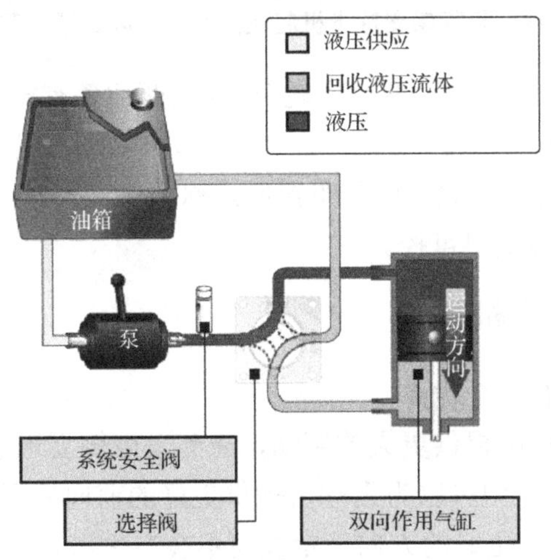

图 4-18　基本液压系统

液压油经系统被油泵输送到作动筒或者伺服机构。伺服机构是一个内部带有活塞的气缸。它将液压能转化为功，产生驱动飞机系统和飞行控制的动力。基于系统的用途，伺服机构可以分为单动式和双动式，即油液可以被输送到伺服机构的单边或双边。因此单动式伺服机构提供单方向的动力。选择阀的作用是控制油液的流向，上述设计使得起落架收放期间，油液流向不同。安全阀为系统提供一个出油口以防液压过大。每个系统结合不同的组成部分来满足不同飞机的需要。

小型飞机上使用最广泛的液压油是矿物基液压油。此类型的液压油是一种类似煤油的石油产品，具有良好的润滑特性，同时也是一种抑制发泡的添加剂，也可阻止腐蚀的形成。它的化学特性非常稳定，黏度随温度变化较小，且被染色，易于识别。通常有好几种类型的液压油可供使用，不同类型飞机应使用制造商指定的液压油类型。具体可以参考飞机飞行手册，飞行员操作手册或者维修手册。

4.9.1 起落架

起落架是飞机在地面上的主要支撑部分。最常见的起落架由轮子构成，但是有的飞机也可以装配用于水上运行的浮子，或者用于雪地上着陆的滑橇，如图 4-19 所示。

图 4-19　起落架支持飞机起飞滑跑、着陆、滑行和停机

小型飞机的起落架有三个轮子：两个主轮(机身的两边各有一个)和另一个位于飞机头部或尾部的轮子。早期，尾轮飞机通常使用第三个轮子安装在飞机尾部的传统起落架。当第三个轮子位于机头位置时，被称为前轮，这种布置形式称为前三点式。可转向的前轮或者尾轮使飞机在地面运行的整个过程变得可控。

4.9.2 前三点式起落架飞机

前三点式起落架飞机有三个优点：

(1) 飞机高速着陆过程中提供更大的制动，避免发生倒立现象(俗称“拿大顶”)。

(2)在起飞、着陆和滑行中可以为飞行员提供更好的前向视野。

(3)地面滑行期间，具有更好的方向稳定性而不会产生滑移。这是由于飞机的重心(CG)位于飞机前轮，使得飞机能够保持直线运动而不会滑移。

前轮要么是手动转向，要么是脚舵转向。手动转向的前轮用线缆或拉杆连接到方向舵，而脚舵转向的前轮则可以自由旋转。在这两种情况下，都需要使用方向舵踏板操纵飞机。然而，脚舵式前轮的飞机需要飞行员将方向舵踏板和制动器结合使用。

4.9.3 后三点式起落架飞机

在后三点式起落架飞机上，两个主轮安装在机身的重心前方，支撑了飞机机身的大部分重量，而较靠后的尾轮作为第三个支撑点。这种安装方式早期用于大型螺旋桨飞机并且适合地面不够坚实的场地(图 4-20)。

图 4-20 后三点式起落架

后三点式起落架的主要缺点是它的重心位于主轮之后，导致方向稳定性差。如果飞机在地面低速滑跑时转向，飞行员就很难控制方向舵，并且飞机重心会落到主轮前方，从而导致飞机失控旋转。

4.9.4 固定式和收放式起落架

起落架可以分为固定式或收放式。固定式起落架固定于机身外，其优点是结构简单，需要维护的地方较少。收放式起落架的设计使飞机具有流线型的特点，当飞机巡航飞行时起落架可以收回到机身内，如图 4-21 所示。

(a)固定式

(b)收放式

图 4-21 固定式和收放式起落架

4.9.5 制动器

飞机制动器位于主轮上，可以通过手动控制或者脚踏板控制。脚踏板可独立操作，因此允许差动制动控制。地面运行期间，差动制动可以作为前轮/尾轮转向的补充使用。

4.10 供 氧 系 统

大多数高空飞行的飞机都配备了不同类型的固定供氧装置。如果飞机没有固定供氧装置，那么飞行期间就必须有可用的便携式供氧装置。便携式供氧装置通常由一个容器、一个调节器、一个面罩出气口和一个压力表构成。飞机氧气通常存储在 1800～2200psi 的高压系统容器中。当氧气罐周围的温度下降时，氧气罐内的气压会下降，因为氧气量不变时，气压变化与温度有关。如果补充氧气罐上的指示压力值降低，并不代表氧气已经耗尽，只是因为氧气罐被安置在飞机上的没有暖气的地方。高压氧气罐在充气之前要标注上容许的 psi 压力容限值(如 1800psi)。氧气罐只能存储纯度不低于 99.5%的航空氧气。工业氧气不能用于呼吸，因为其中可能含有杂质。而医用氧气中包含水蒸气，当其处于低温环境时，水蒸气可能在调节器中结冰。因此为保证安全，应该定期检查和维护供氧系统。

供氧系统由面罩和根据机舱高度供应氧气流的调节器组成。飞行高度为 18000ft 以上时不允许使用面罩。核准用于高度达 40000ft 的调节器被设计用于在机舱高度为 8000ft 以下时提供 0%气缸氧气和 100%机舱空气，在飞行高度大约 34000ft 时比率改变为 100%氧气和 0%机舱空气。核准用于高度达 45000ft 的调节器被设计用于在低高度时提供 40%气缸氧气和 60%机舱空气，在较高的高度时比率变化到 100%。没有氧气时，飞行员应该避免白天高于 10000ft 的高度飞行，避免夜晚高于 8000ft 的高度飞行(图 4-22)。

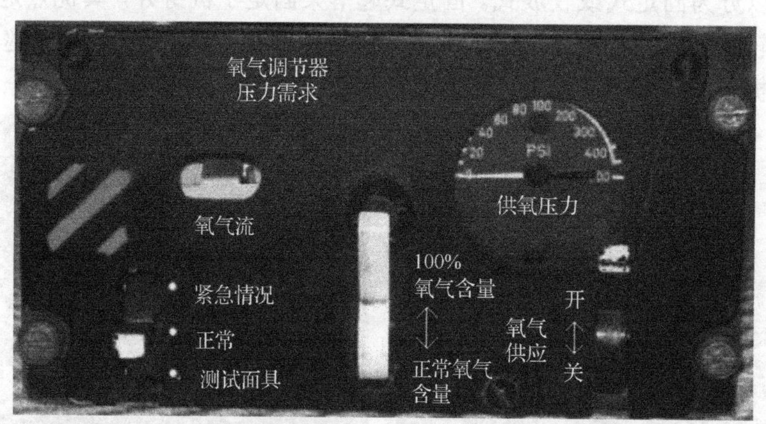

图 4-22　氧气系统调节器

飞行员在使用氧气时，应该意识到火灾的危险性。通常耐火性差的材料在氧气中容易燃烧。润滑油或者油脂如果暴露在氧气中可能着火，因此它们不能用于氧气设备的配件和阀门的密封。使用任何类型的氧气设备期间都禁止吸烟。每次飞行之前，飞行员应该全面检查和测试供氧设备。检查时全面测试飞机氧气设备，包括氧气可用量、系统工作状态、确保备用氧气设备随时可用。检查时保持双手干净，并目视检查面罩、管道是否存在撕裂、破裂或变质；检查调节阀门和调节杆的状态和位置；检查氧气含量；检查氧气压力表的位置和运行状态、检查流量指示器和连接状态。测试系统时应该带上氧气面罩。任何时候使用氧气后，都要确认所有部件和阀门都已关闭。

4.11 防冰系统

防冰装置用于阻止冰的形成，而除冰装置用于除掉已经形成的冰。这些系统能够保护机翼和尾翼面的前缘、空速管和静压口的开口、油箱通风管、失速告警装置、挡风板和螺旋桨叶片。一些飞机上安装了结冰检测灯，用来检测夜晚飞行时的结构性结冰的强度。

4.11.1 机翼防冰和除冰

可膨胀的除冰罩由搭接到机翼前缘的橡胶薄板组成。当冰在前缘形成时，由发动机驱动气动泵使橡胶罩膨胀。一些涡轮螺旋桨飞机把发动机的排气转向到机翼，使橡胶罩膨胀。一旦膨胀，冰就会破碎，并从机翼前缘掉落。除冰罩由驾驶舱内的开关控制，可以单次运行，也可以定期自动运行(图 4-23)。

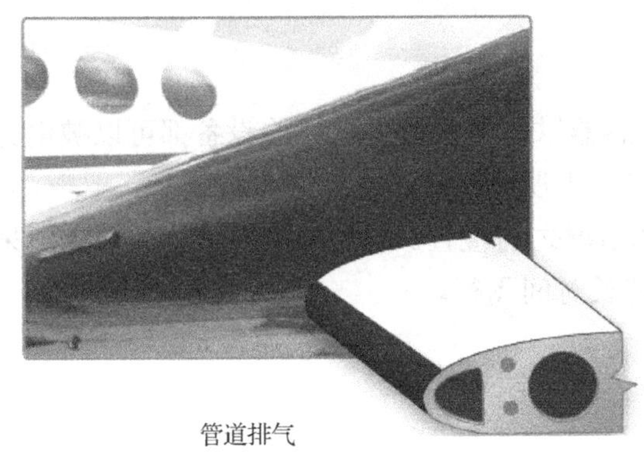
管道排气

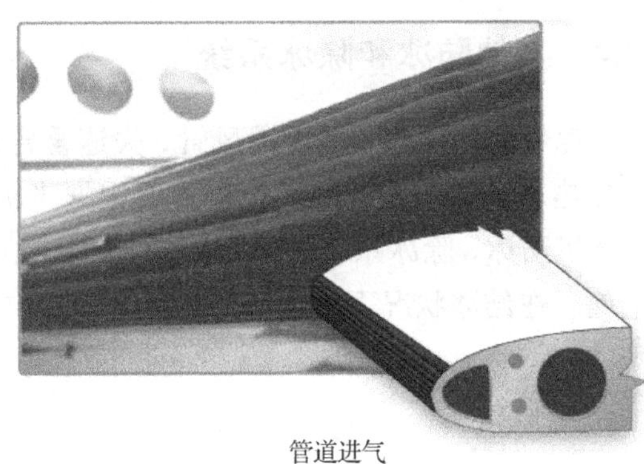
管道进气

图 4-23 机翼前段除冰罩

很多除冰罩系统使用仪表系统的真空计和气压计来指示合适的除冰罩操作。这些仪表有指示除冰操作限制范围的标记。一些系统也会配备信号灯来指示正确的除冰操作。

除冰罩的维护和保养对防冰系统的持续运行起重要作用，因此在飞行前应该仔细检查除冰罩。

另一个机翼前缘保护方式是热防冰系统。因为热量是防止机翼结冰的有效方法之一，它是通过高性能涡轮飞机的发动机压缩段的热空气导入到翼面前缘来实现的。这个系统在结冰前开始运作。热空气加热前缘，能够有效地阻止结冰。

可选类型的前缘保护不像热防冰和除冰罩那样常见，它被称为渗漏机翼。渗漏机翼有一个位于机翼前缘的小洞，以防止冰的形成和积聚，具体操作是将防冻液泵送到前缘并通过这个小洞排出。

4.11.2 挡风玻璃防冰

挡风玻璃防冰系统主要有两种方法：一种方法是系统引导酒精流到挡风玻璃上。如果酒精使用时间够早，就可以阻止冰堆积在挡风玻璃上。酒精流的速度可以根据飞机制造商建议的程序并通过驾驶舱中的刻度盘来控制。

另一种有效方法是电加热法，即将小金属丝或者其他导电材料嵌入挡风玻璃。通过驾驶舱中的开关打开加热器，此时电流通过屏蔽层的金属丝提供足够的热量来防止挡风玻璃结冰。受热的挡风玻璃只能在飞行时使用。地面运行时禁止打开加热器，因为这会导致过热而使得挡风玻璃损坏。注意：电流会使罗盘偏转误差达 40°。

4.11.3　螺旋桨防冰

螺旋桨通过使用酒精或者电加热元件来防止结冰。一些螺旋桨配备了指向桨叶根部的排放喷嘴。酒精从喷嘴放出，离心力使酒精流向桨叶的前缘来防止螺旋桨的前缘结冰。螺旋桨上也可以安装螺旋桨除冰罩。螺旋桨除冰罩分为两部分：舱内和舱外。除冰罩经过开槽有利于酒精的导流，并且还嵌入了带有电流的电线用于加热螺旋桨。螺旋桨除冰系统的运行状态可以通过螺旋桨防冰电流表查看。飞行前检查期间要检查螺旋桨除冰罩的正常运行。如果除冰罩不能加热某个桨叶，则会导致不均衡的桨叶载荷，进而引起严重的螺旋桨振动。

4.11.4　其他防冰和除冰系统

空速管和静压口、燃油通风管、失速警告传感器以及其他可供选择的设备都可以被电力装置加热。电加热系统的运行检查要根据飞机飞行手册或飞行员操作手册来进行。

飞机防冰、除冰系统的运行应该在遇到结冰情况之前检查。遇到结构性结冰时要立即采取措施。在结冰状况下防冰和除冰设备无法维持长时间飞行。

Chapter 5 Flight Instruments

In order to safely fly any aircraft, a pilot must understand how to interpret and operate the flight instruments. The pilot also needs to be able to recognize associated errors and malfunctions of these instruments. This chapter addresses the Pitot-static system and associated instruments, the vacuum system and related instruments, gyroscopic instruments, and the magnetic compass. When a pilot understands how each instrument works and recognizes when an instrument is malfunctioning, he or she can safely utilize the instruments to their fullest potential.

5.1 Pitot-Static Flight Instruments

The Pitot-static system is a combined system that utilizes the static air pressure, and the dynamic pressure due to the motion of the aircraft through the air. These combined pressures are utilized for the operation of the airspeed indicator(ASI), altimeter, and vertical speed indicator (VSI)(Figure 5-1).

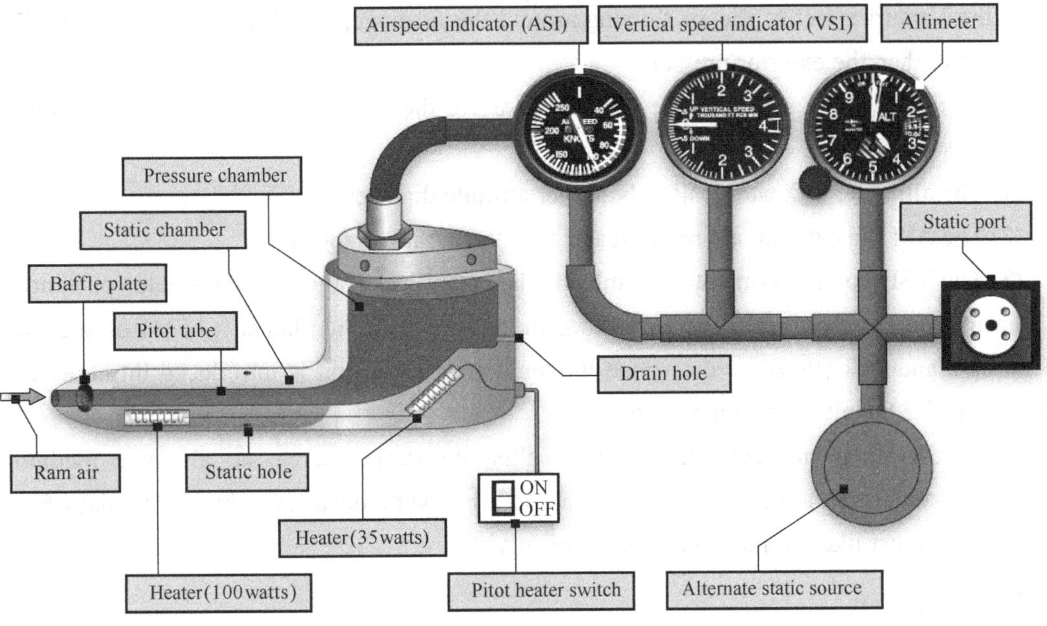

Figure 5-1 Pitot-static system and instruments

5.1.1 Impact Pressure Chamber and Lines

The Pitot tube is utilized to measure the total combined pressures that are present when an aircraft moves through the air. Static pressure, also known as ambient pressure, is always present whether an aircraft is moving or at rest. It is simply the barometric pressure in the local area. Dynamic pressure is present only when an aircraft is in motion; therefore, it can be thought of as a pressure due to motion. Wind also generates dynamic pressure. It does not matter if the aircraft is moving through still air at 70 knots or if the aircraft is facing a wind with a speed of 70 knots, the same dynamic pressure is generated.

When the wind blows from an angle less than 90° off the nose of the aircraft, dynamic pressure can be depicted on the ASI. The wind moving across the airfoil at 20 knots is the same as the aircraft moving through calm air at 20 knots. The Pitot tube captures the dynamic pressure, as well as the static pressure that is always present. The Pitot tube has a small opening at the front which allows the total pressure to enter the pressure chamber. The total pressure is made up of dynamic pressure plus static pressure.

5.1.2 Static Chamber and Lines

The static chamber is vented through small holes to the free undisturbed air on the side(s) of the aircraft. As the atmospheric pressure changes, the pressure is able to move freely in and out of the instruments through the small lines which connect the instruments into the static system. Should the primary static source become blocked, an alternate static source is provided in some aircrafts to provide static pressure. The alternate static source is normally found inside of the flight deck. Due to the Venturi effect of the air flowing around the fuselage, the air pressure inside the flight deck is lower than the exterior pressure.

When the alternate static source pressure is used, the following instrument indications are observed:

(1) The altimeter indicates a slightly higher altitude than actual.

(2) The ASI indicates an airspeed greater than the actual airspeed.

(3) The VSI shows a momentary climb and then stabilizes if the altitude is held constant.

Each pilot is responsible for consulting the Aircraft Flight Manual (AFM) or the Pilot's Operating Handbook (POH) to determine the amount of error that is introduced into the system when we utilize the alternate static source.

If an aircraft is not equipped with an alternate static source, an alternate method of introducing static pressure into the system should a blockage occur is to break the glass face of the VSI. This most likely renders the VSI inoperative.

5.1.3 Altimeter

The altimeter is an instrument that measures the height of an aircraft above a given pressure level. Since the altimeter is the only instrument that is capable of indicating altitude, this is one of the most vital instruments installed in the aircraft. To use the altimeter effectively, the pilot must understand the operation of the instrument, as well as the errors associated with the altimeter and how each effects the indication.

A stack of sealed aneroid wafers comprise the main component of the altimeter. A mechanical linkage connects the wafer movement to the needles on the indicator face, which translates compression of the wafers into a decrease in altitude and translates an expansion of the wafers into an increase in altitude (Figure 5-2).

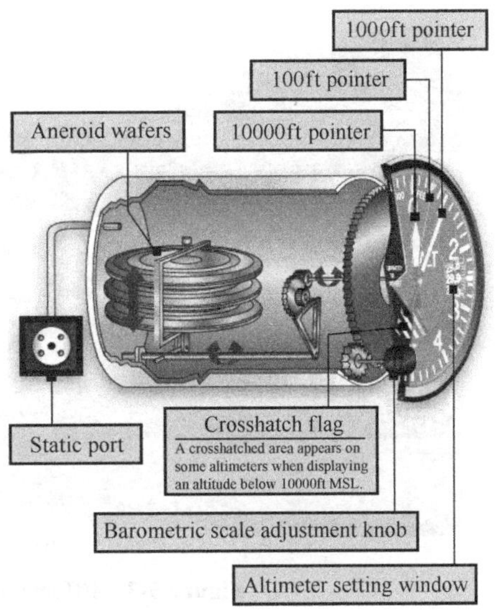

Figure 5-2 Altimeter

1. Principle of Operation

The pressure altimeter is an aneroid barometer that measures the pressure of the atmosphere at the level where the altimeter is located, and presents an altitude indication in feet. The altimeter uses static pressure as its source of operation. Air is denser at sea level than aloft—as altitude increases, atmospheric pressure decreases. This difference in pressure at various levels causes the altimeter to indicate changes in altitude.

The presentation of altitude varies considerably between different types of altimeters. Some have one pointer while others have two or more. Only the multipointer type is discussed in this book. The dial of a typical altimeter is graduated with numerals arranged clockwise from zero to nine. Movement of the aneroid element is transmitted through gears to the three hands that indicate altitude. The shortest hand indicates altitude in tens of thousands of feet, the intermediate hand in thousands of feet, and the longest hand in hundreds of feet.

This indicated altitude is correct, however, only when the sea level barometric pressure is standard (29.92in Hg), the sea level free air temperature is standard +15 degrees Celsius (℃) or 59 degrees Fahrenheit (℉), and the pressure and temperature decrease at a standard rate with an increase in altitude. Adjustments for nonstandard pressures are accomplished by setting the corrected pressure into a barometric scale located on the face of the altimeter. The barometric pressure window is sometimes referred to as the Kollsman window; only after the altimeter is set does it indicate the correct altitude.

2. Effect of Nonstandard Pressure and Temperature

If altimeters could not be adjusted for nonstandard pressure, a hazardous situation could occur. For example, if an aircraft is flown from a high pressure area to a low pressure area without adjusting the altimeter, a constant altitude will be displayed, but the actual height of the aircraft above the ground would be lower than the indicated altitude. There is an old aviation axiom: "GOING FROM A HIGH TO A LOW, LOOK OUT BELOW". Conversely, if an aircraft is flown from a low pressure area to a high pressure area without an adjustment of the altimeter, the actual altitude of the aircraft is higher than the indicated altitude (Figure 5-3).

Many altimeters do not have an accurate means of being adjusted for barometric pressures in excess of 31.00 inches of mercury (in Hg). When the altimeter cannot be set to the higher pressure setting, the aircraft actual altitude will be higher than that the altimeter indicates. When low

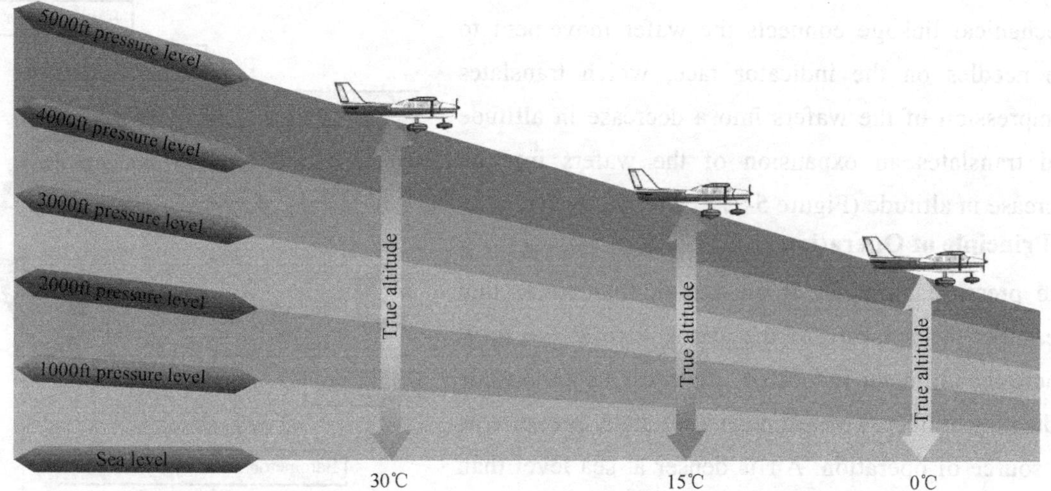

Figure 5-3　Effects of nonstandard temperature on an altimeter

barometric pressure conditions occur (below 28.00), flight operations by aircraft unable to set the actual altimeter setting are not recommended. If terrain or obstacle clearance is a factor in selecting a cruising altitude, particularly in mountainous terrain, remember to anticipate that a colder-than-standard temperature places the aircraft lower than that the altimeter indicates. Therefore, a higher indicated altitude may be required to provide adequate terrain clearance. A variation of the memory aid used for pressure can be employed: "FROM HOT TO COLD, LOOK OUT BELOW."

3. Setting the Altimeter

Most altimeters are equipped with a barometric pressure setting window (or Kollsman window) providing a means to adjust the altimeter. A knob is located at the bottom of the instrument for this adjustment.

To adjust the altimeter for variation in atmospheric pressure, the pressure scale in the altimeter setting window, calibrated in inches of mercury (in Hg) and/or millibars (mbar), is adjusted to match the given altimeter setting. Altimeter setting is defined as station pressure reduced to sea level, but, an altimeter setting is accurate only in the vicinity of the reporting station. Therefore, the altimeter must be adjusted as the flight progresses from one station to the next.

Many pilots confidently expect the current altimeter setting will compensate for irregularities in atmospheric pressure at all altitudes, but this is not always true. The altimeter setting broadcast by ground stations is the station pressure corrected to mean sea level. It does not account for the irregularities at higher levels, particularly the effect of nonstandard temperature. If each pilot in a given area is using the same altimeter setting, each altimeter should be equally affected by temperature and pressure variation errors, making it possible to maintain the desired vertical separation between aircraft. When pilots fly over high, mountainous terrain, certain atmospheric conditions cause the altimeter to indicate an altitude of 1000ft or more higher than the actual altitude. For this reason, a generous margin of altitude should be allowed—not only for possible altimeter error, but also for possible downdrafts that might be associated with high winds.

To illustrate the use of the altimeter setting system, follow a flight from Dallas Love Field, Texas, to Abilene Municipal Airport, Texas, via Mineral Wells. Before taking off from Love Field, the pilot receives a current altimeter setting of 29.85 in Hg from the control tower or ATIS, and sets this value in the altimeter setting window. The altimeter indication should then be compared with the known airport elevation of 487ft. Since most altimeters are not perfectly calibrated, an error may exist.

When over Mineral Wells, assume the pilot receives a current altimeter setting of 29.94 in Hg and sets this in the altimeter window. Before pilots enter the traffic pattern at Abilene Municipal Airport, a new altimeter setting of 29.69 in Hg is received from the Abilene Control Tower, and set in the altimeter setting window. If the pilot desires to fly the traffic pattern at approximately 800ft above the terrain, and the field elevation of Abilene is 1791ft, an indicated altitude of 2600ft should be maintained (1791ft + 800ft = 2591ft, rounded to 2600ft).

The importance of properly setting the altimeter cannot be overemphasized. Assume the pilot did not adjust the altimeter at Abilene to the current setting and continued using the Mineral Wells setting of 29.94 in Hg. When entering the Abilene traffic pattern at an indicated altitude of 2600ft, the aircraft would be approximately 250ft below the proper traffic pattern altitude. Upon landing, the altimeter would indicate approximately 250ft higher than the field elevation.

(1) Mineral Wells altimeter setting 29.94.

(2) Abilene altimeter setting 29.69.

(3) Difference 0.25.

(Since 1 inch of pressure is equal to approximately 1000ft of altitude, 0.25×1000ft = 250ft.)

4. Altimeter Operation

There are two means by which the altimeter pointers can be moved. The first is a change in air pressure, while the other is an adjustment to the barometric scale. When the aircraft climbs or descends, changing pressure within the altimeter case expands or contracts the aneroid barometer. This movement is transmitted through mechanical linkage to rotate the pointers.

A decrease in pressure causes the altimeter to indicate an increase in altitude, and an increase in pressure causes the altimeter to indicate a decrease in altitude. Knowing the aircraft's altitude is vitally important to a pilot. The pilot must be sure that the aircraft is flying high enough to clear the highest terrain or obstruction along the intended route. It is especially important to have accurate altitude information when visibility is restricted. To clear obstructions, the pilot must constantly be aware of the altitude of the aircraft and the elevation of the surrounding terrain. To reduce the possibility of a midair collision, it is essential to maintain altitude in accordance with air traffic rules.

5. Types of Altitude

Altitude is vertical distance above some point or level used as a reference. There are as many kinds of altitude as there are reference levels from which altitude is measured, and each may be used for specific reasons. Pilots are mainly concerned with five types of altitudes:

(1) Indicated altitude—read directly from the altimeter (uncorrected) when it is set to the current altimeter setting.

(2) True altitude—the vertical distance of the aircraft above sea level—the actual altitude. It is often expressed as feet above mean sea level (MSL). Airport, terrain, and obstacle elevations on aeronautical charts are true altitudes.

(3) Absolute altitude—the vertical distance of an aircraft above the terrain, or above ground level (AGL).

(4) Pressure altitude—the altitude indicated when the altimeter setting window (barometric scale) is adjusted to 29.92in Hg. This is the altitude above the standard datum plane, which is a theoretical plane where air pressure (corrected to 15℃) equals 29.92in Hg. Pressure altitude is used to compute density altitude, true altitude, true airspeed (TAS), and other performance data.

(5) Density altitude—pressure altitude corrected for variations from standard temperature. When conditions are standard, pressure altitude and density altitude are the same. If the temperature is above standard, the density altitude is higher than pressure altitude. If the temperature is below standard, the density altitude is lower than pressure altitude. This is an important altitude because it is directly related to the aircraft's performance.

As an example, consider an airport with a field elevation of 5048ft MSL where the standard temperature is 5℃. Under these conditions, pressure altitude and density altitude are the same—5048ft. If the temperature changes to 30℃, the density altitude increases to 7855ft. This means an aircraft would perform on takeoff as though the field elevation were 7855ft at standard temperature. Conversely, a temperature of −25℃ would result in a density altitude of 1232ft. An aircraft would perform much better under these conditions.

6. Instrument Check

To determine the condition of an altimeter, set the barometric scale to the current reported altimeter setting transmitted by the local automated flight service station (AFSS) or any other reliable source, such as ATIS, AWOS, or ASOS. The altimeter pointers should indicate the surveyed field elevation of the airport. If the indication is off more than 75ft from the surveyed field elevation, the instrument should be referred to a certificated instrument repair station for recalibration.

5.1.4 Vertical Speed Indicator

The VSI, which is sometimes called a vertical velocity indicator (VVI), indicates whether the aircraft is climbing, descending, or in level flight. The rate of climb or descent is indicated in feet per minute (fpm). If properly calibrated, the VSI indicates zero in level flight (Figure 5-4).

Although the VSI operates solely from static pressure, it is a differential pressure instrument. It contains a diaphragm with connecting linkage and gearing to the indicator pointer inside an airtight case. The inside of the diaphragm is connected directly to the static line of the Pitot-static system. The area outside the diaphragm, which is inside the instrument case, is also connected to the static line, but through a restricted orifice (calibrated leak).

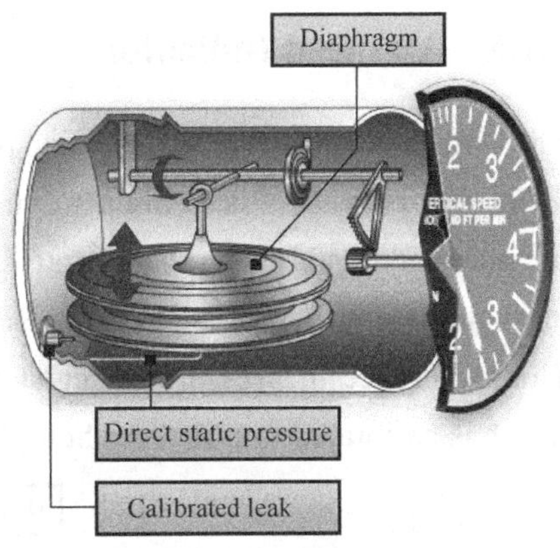

Figure 5-4 Vertical speed indicator

Both the diaphragm and the case receive air from the static line at existing atmospheric pressure. The diaphragm receives unrestricted air while the case receives the static pressure via the metered leak. When the aircraft is on the ground or in level flight, the pressures inside the diaphragm and the instrument case are equal and the pointer is at the zero indication. When the aircraft climbs or descends, the pressure inside the diaphragm changes immediately, but due to the metering action of the restricted passage, the case pressure remains higher or lower for a short time, causing the diaphragm to contract or expand. This causes a pressure differential that is indicated on the instrument needle as a climb or descent. When the pressure differential stabilizes at a definite ratio, the needle indicates the rate of altitude change.

The VSI displays two different types of information:

(1) Trend information shows an immediate indication of an increase or decrease in the aircraft's rate of climb or descent.

(2) Rate information shows a stabilized rate of change in altitude.

The time period from the initial change in the rate of climb, until the VSI displays an accurate indication of the new rate, is called the lag. Rough control technique and turbulence can extend the lag period and cause erratic and unstable rate indications. Some aircraft are equipped with an instantaneous vertical speed indicator (IVSI), which incorporates accelerometers to compensate for the lag in the typical VSI (Figure 5-5).

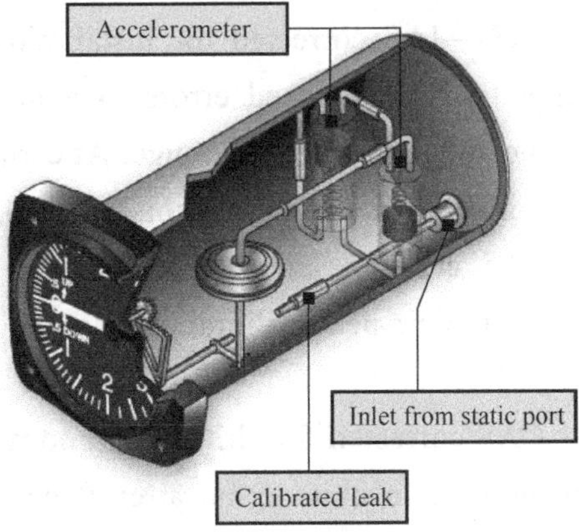

Figure 5-5 An IVSI incorporates accelerometers to help the instrument immediately indicate changes in vertical speed

5.1.5 Airspeed Indicator

The ASI is a sensitive, differential pressure gauge which measures and promptly indicates the difference between Pitot (impact/dynamic pressure) and static pressure (Figure 5-6). These two pressures are equal when the aircraft is parked on the ground in calm air. When the aircraft moves through the air, the pressure on the Pitot line becomes greater than the pressure in the static lines. This difference in pressure is registered by the airspeed pointer on the face of the instrument, which is calibrated in miles per hour, knots (nautical miles per hour), or both.

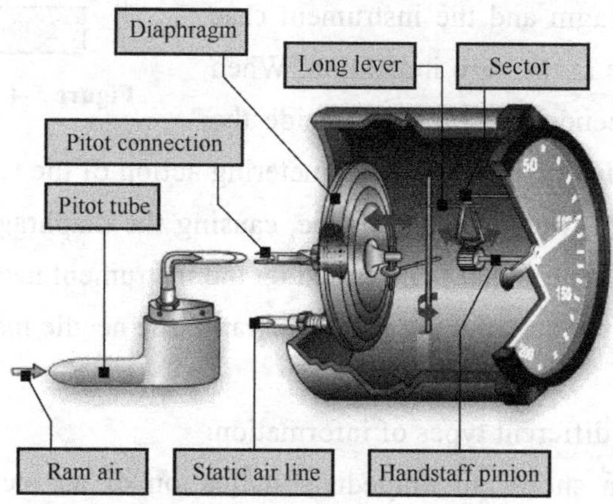

Figure 5-6　Airspeed indicator

Pilots need to be very familiar with each type.

(1) Indicated airspeed (IAS)—the direct instrument reading obtained from the ASI, uncorrected for variations in atmospheric density, installation error, or instrument error. Manufacturers use this airspeed as the basis for determining aircraft performance. Takeoff, landing, and stall speeds listed in the AFM/ POH are IAS and do not normally vary with altitude or temperature.

(2) Calibrated airspeed (CAS)—IAS corrected for installation error and instrument error. Although manufacturers attempt to keep airspeed errors to a minimum, it is not possible to eliminate all errors throughout the airspeed operating range. At certain airspeeds and with certain flap settings, the installation and instrument errors may total several knots. This error is generally greatest at low airspeeds. In the cruising and higher airspeed ranges, IAS and CAS are approximately the same. Refer to the airspeed calibration chart to correct for possible airspeed errors.

(3) True airspeed (TAS)—CAS corrected for altitude and nonstandard temperature. Because air density decreases with an increase in altitude, an aircraft has to be flown faster at higher altitudes to cause the same pressure difference between Pitot impact pressure and static pressure. Therefore, for a given CAS, TAS increases as altitude increases; or for a given TAS, CAS decreases as altitude increases. A pilot can find TAS by two methods. The most accurate method

is to use a flight computer. With this method, the CAS is corrected for temperature and pressure variation by using the airspeed correction scale on the computer. Extremely accurate electronic flight computers are also available. Just enter the CAS, pressure altitude, and temperature, and the computer calculates the TAS. A second method, which is a rule of thumb, provides the approximate TAS. Simply add 2 percent to the CAS for each 1000ft of altitude. The TAS is the speed which is used for flight planning and is used when filing a flight plan.

(4) Groundspeed (GS)—the actual speed of the airplane over the ground. It is TAS adjusted for wind. GS decreases with a headwind, and increases with a tailwind.

1. Airspeed Indicator Markings

Aircraft weighing 12500 pounds or less, manufactured after 1945, and certificated by the FAA, are required to have ASIs marked in accordance with a standard color-coded marking system. This system of color-coded markings enables a pilot to determine at a glance certain airspeed limitations that are important to the safe operation of the aircraft. For example, if during the execution of a maneuver, it is noted that the airspeed needle is in the yellow arc and rapidly approaching the red line, the immediate reaction should be to reduce airspeed.

As shown in Figure 5-7, ASIs on single-engine small aircraft include the following standard color-coded markings:

(1) White arc—commonly referred to as the flap operating range since its lower limit represents the full flap stall speed and its upper limit provides the maximum flap speed. Approaches and landings are usually flown at speeds within the white arc.

(2) Lower limit of white arc (V_{SO})—the stalling speed or the minimum steady flight speed in the landing configuration. In small aircraft, this is the power-off stall speed at the maximum landing weight in the landing configuration (gear and flaps down).

(3) Upper limit of the white arc (V_{FE})—the maximum speed with the flaps extended.

(4) Green arc—the normal operating range of the aircraft. Most flying occurs within this range.

(5) Lower limit of green arc (V_{SI})—the stalling speed or the minimum steady flight speed obtained in a specified configuration. For most aircraft, this is the power-off stall speed at the maximum takeoff weight in the clean configuration (gear up, if retractable, and flaps up).

(6) Upper limit of green arc (V_{NO})—the maximum structural cruising speed. Do not exceed this speed except in smooth air.

(7) Yellow arc—caution range. Fly within this range only in smooth air, and then, only with caution.

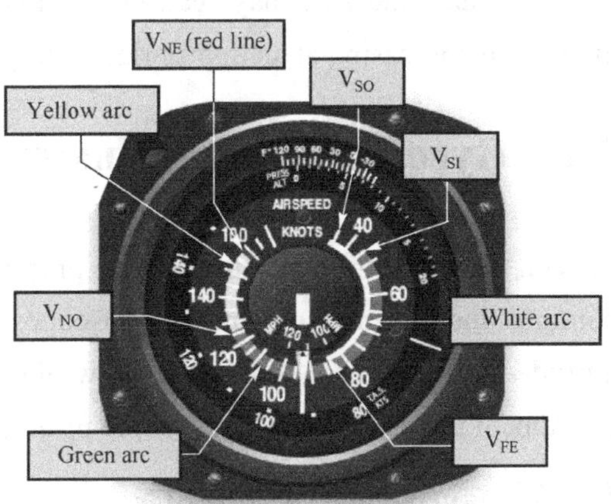

Figure 5-7 Airspeed indicator markings

(8) Red line (V_{NE})—never exceed speed. Operating above this speed is prohibited since it may result in damage or structural failure.

2. Other Airspeed Limitations

Some important airspeed limitations are not marked on the face of the ASI, but are found on placards and in the AFM/POH. These airspeeds include:

(1) Design maneuvering speed (V_A)—the maximum speed at which the structural design's limit load can be imposed (either by gusts or full deflection of the control surfaces) without causing structural damage. It is important to consider weight when we reference this speed. For example, VA may be 100 knots when an airplane is heavily loaded, but only 90 knots when the load is light.

(2) Landing gear operating speed (V_{LO})—the maximum speed for extending or retracting the landing gear if pilots fly an aircraft with retractable landing gear.

(3) Landing gear extended speed (V_{LE})—the maximum speed at which an aircraft can be safely flown with the landing gear extended.

(4) Best angle-of-climb speed (V_X)—the airspeed at which an aircraft gains the greatest amount of altitude in a given distance. It is used during a short-field takeoff to clear an obstacle.

(5) Best rate-of-climb speed (V_Y)—the airspeed that provides the most altitude gain in a given period of time.

(6) Single-engine best rate-of-climb (V_{YSE})—the best rate-of-climb or minimum rate-of-sink in a light twin-engine aircraft with one engine inoperative. It is marked on the ASI with a blue line. V_{YSE} is commonly referred to as "Blue Line".

(7) Minimum control speed (V_{MC})—the minimum flight speed at which a light, twin-engine aircraft can be satisfactorily controlled when an engine suddenly becomes inoperative and the remaining engine is at takeoff power.

3. Instrument Check

Prior to takeoff, the ASI should read zero. However, if there is a strong wind blowing directly into the Pitot tube, the ASI may read higher than zero. When beginning the takeoff, make sure the airspeed is increasing at an appropriate rate.

5.2 Gyroscopic Flight Instruments

Several flight instruments utilize the properties of a gyroscope for their operation. The most common instruments containing gyroscopes are the turn coordinator, heading indicator, and the attitude indicator. To understand how these instruments operate requires knowledge of the instrument power systems, gyroscopic principles, and the operating principles of each instrument.

5.2.1 Gyroscopic Principles

Any spinning object exhibits gyroscopic properties. A wheel or rotor designed and mounted to utilize these properties is called a gyroscope. Two important design characteristics of an instrument gyro are great weight for its size, or high density, and rotation at high speed with low friction bearings.

There are two general types of mountings; the type used depends upon which property of the gyro is utilized. A freely or universally mounted gyroscope is free to rotate in any direction about its center of gravity. Such a wheel is said to have three planes of freedom. The wheel or rotor is free to rotate in any plane in relation to the base and is balanced so that, with the gyro wheel at rest, it remains in the position in which it is placed. Restricted or semi-rigidly mounted gyroscopes are those mounted so that one of the planes of freedom is held fixed in relation to the base.

There are two fundamental properties of gyroscopic action: rigidity in space and precession.

1. Rigidity in Space

Rigidity in space refers to the principle that a gyroscope remains in a fixed position in the plane in which it is spinning (Figure 5-8).

By mounting this wheel, or gyroscope, on a set of gimbal rings, the gyro is able to rotate freely in any direction. Thus, if the gimbal rings are tilted, twisted, or otherwise moved, the gyro remains in the plane in which it was originally spinning.

2. Precession

Precession is the tilting or turning of a gyro in response to a deflective force. The reaction to this force does not occur at the point at which it was applied; rather, it occurs at a point that is 90° later in the direction of rotation. This principle allows the gyro to determine a rate of turn by sensing the amount of pressure created by a change in direction. The rate at which the gyro precesses is inversely proportional to the speed of the rotor and proportional to the deflective force (Figure 5-9).

Figure 5-8 Regardless of the position of its base, a gyro tends to remain rigid in space, with its axis of rotation pointed in a constant direction

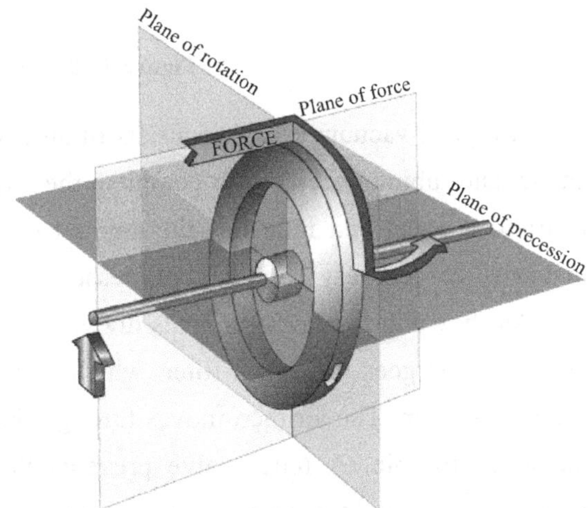

Figure 5-9 Precession of a gyroscope resulting from an applied deflective force

Precession can also create some minor errors in some instruments.

3. Sources of Power

In some aircraft, all the gyros are vacuum, pressure, or electrically operated. In other aircraft, vacuum or pressure systems provide the power for the heading and attitude indicators, while the electrical system provides the power for the turn coordinator. Most aircraft have at least two sources of power to ensure at least one source of bank information is available if one power source fails. The vacuum or pressure system spins the gyro by drawing a stream of air against the rotor vanes to spin the rotor at high speed, much like the operation of a waterwheel or turbine. The amount of vacuum or pressure required for instrument operation varies, but is usually between 4.5in Hg and 5.5in Hg.

One source of vacuum for the gyros is a vane-type engine-driven pump that is mounted on the accessory case of the engine. Pump capacity varies in different airplanes, depending on the number of gyros (Figure 5-10).

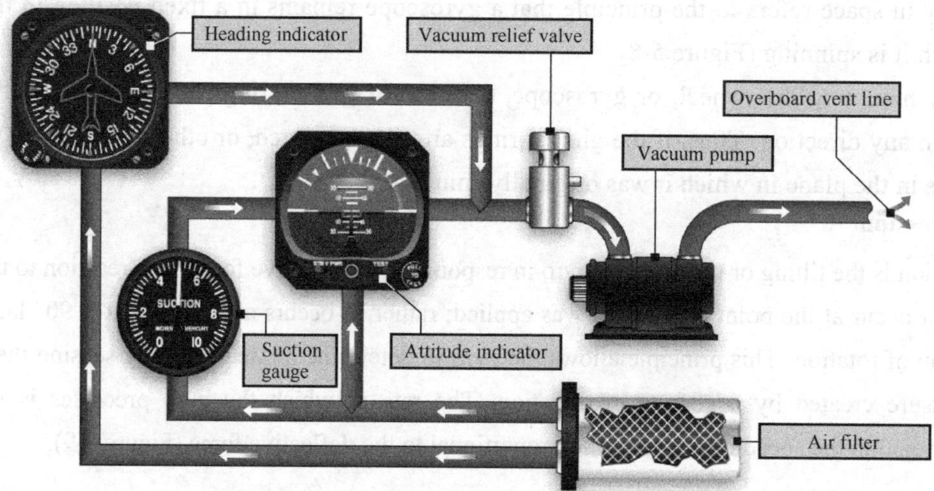

Figure 5-10　Typical vacuum system

A typical vacuum system consists of an engine-driven vacuum pump, relief valve, air filter, gauge, and tubing necessary to complete the connections. The gauge is mounted in the aircraft's instrument panel and indicates the amount of pressure in the system (vacuum is measured in inches of mercury less than ambient pressure).

As shown in Figure 5-10, air is drawn into the vacuum system by the engine-driven vacuum pump. It first goes through a filter, which prevents foreign matter from entering the vacuum or pressure system. The air then moves through the attitude and heading indicators, where it causes the gyros to spin. A relief valve prevents the vacuum pressure, or suction, from exceeding prescribed limits. After that, the air is expelled overboard or used in other systems, such as for inflating pneumatic deicing boots.

It is important to monitor vacuum pressure during flight, because the attitude and heading indicators may not provide reliable information when suction pressure is low. The vacuum, or suction, gauge is generally marked to indicate the normal range. Some aircraft are equipped with a warning light that illuminates when the vacuum pressure drops below the acceptable level.

5.2.2 Turn Indicators

Aircraft use two types of turn indicators: turn-and-slip indicator and turn coordinator. Because of the way the gyro is mounted, the turn-and-slip indicator shows only the rate of turn in degrees per second. The turn coordinator is mounted at an angle, or canted, so it can initially show roll rate. When the roll stabilizes, it indicates rate of turn. Both instruments indicate turn direction and quality (coordination), and also serve as a backup source of bank information in the event an attitude indicator fails. Coordination is achieved by referring to the inclinometer, which consists of a liquid-filled curved tube with a ball inside (Figure 5-11).

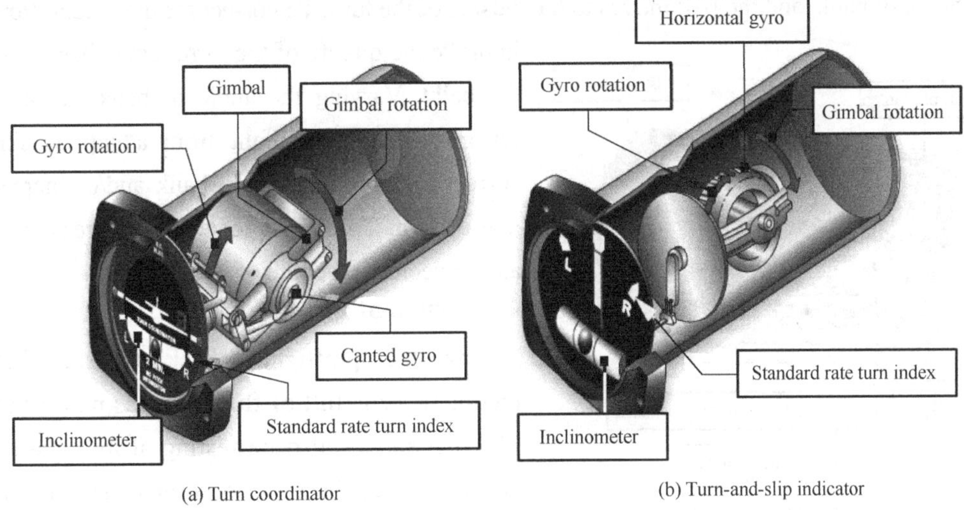

(a) Turn coordinator (b) Turn-and-slip indicator

Figure 5-11 Turn indicators rely on controlled precession for their operation

1. Turn-and-Slip Indicator

The gyro in the turn-and-slip indicator rotates in the vertical plane, corresponding to the aircraft's longitudinal axis. A single gimbal limits the planes in which the gyro can tilt, and a spring tries to return it to center. Because of precession, a yawing force causes the gyro to tilt left or right, as viewed from the pilot seat. The turn-and-slip indicator uses a pointer, called the turn needle, to show the direction and rate of turn.

2. Turn Coordinator

The gimbal in the turn coordinator is canted; therefore, its gyro can sense both rate of roll and rate of turn. Since turn coordinators are more prevalent in training aircraft, this discussion concentrates on that instrument. When rolling into or out of a turn, the miniature aircraft banks in the direction the aircraft is rolled. A rapid roll rate causes the miniature aircraft to bank more steeply than a slow roll rate.

The turn coordinator can be used to establish and maintain a standard-rate turn by aligning the wing of the miniature aircraft with the turn index.

3. Inclinometer

The inclinometer is used to depict aircraft yaw, which is the side-to-side movement of the aircraft's nose. During coordinated, straight-and-level flight, the force of gravity causes the ball to rest in the lowest part of the tube, centered between the reference lines. Coordinated flight is maintained by keeping the ball centered. If the ball is not centered, it can be centered by using the rudder.

To center the ball, apply rudder pressure on the side to which the ball is deflected. Use the simple rule, "step on the ball", to remember which rudder pedal to press. If aileron and rudder are coordinated during a turn, the ball remains centered in the tube. If aerodynamic forces are unbalanced, the ball moves away from the center of the tube. As shown in Figure 5-12, in a slip, the rate of turn is too slow for the angle of bank, and the ball moves to the inside of the turn. In a skid, the rate of turn is too great for the angle of bank, and the ball moves to the outside of the turn. To correct for these conditions, and improve the quality of the turn, remember to "step on the ball". Varying the angle of bank can also help restore coordinated flight from a slip or skid. To correct for a slip, decrease bank and/or increase the rate of turn. To correct for a skid, increase the bank and/or decrease the rate of turn.

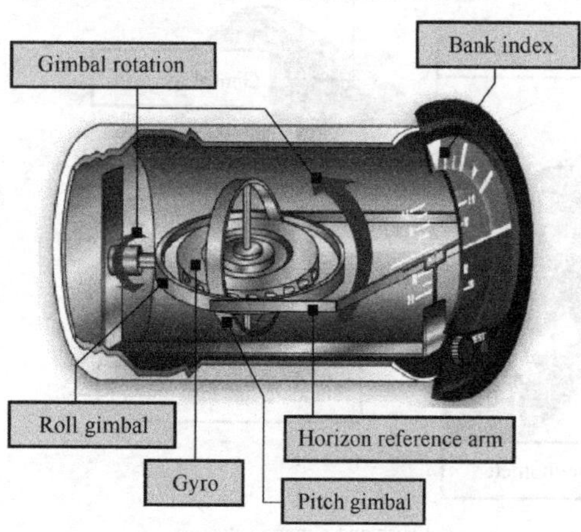

Figure 5-12　Attitude indicator

4. Instrument Check

During the preflight, check to see that the inclinometer is full of fluid and has no air bubbles. The ball should also be resting at its lowest point. When taxiing, the turn coordinator should indicate a turn in the correct direction while the ball moves opposite the direction of the turn.

5.2.3　Attitude Indicator

The attitude indicator, with its miniature aircraft and horizon bar, displays a picture of the attitude of the aircraft. The relationship of the miniature aircraft to the horizon bar is the same as the relationship of the real aircraft to the actual horizon. The instrument gives an instantaneous indication of even the smallest changes in attitude.

The gyro in the attitude indicator is mounted in a horizontal plane and depends upon rigidity in space for its operation. The horizon bar represents the true horizon. This bar is fixed to the gyro and remains in a horizontal plane as the aircraft is pitched or banked about its lateral or longitudinal axis, indicating the attitude of the aircraft relative to the true horizon.

An adjustment knob is provided with which the pilot may move the miniature aircraft up or down to align the miniature aircraft with the horizon bar to suit the pilot's line of vision. Normally, the miniature aircraft is adjusted so that the wings overlap the horizon bar when the aircraft is in straight-and-level cruising flight (Figure 5-12).

The pitch and bank limits depend upon the make and model of the instrument. Limits in the banking plane are usually from 100° to 110°, and the pitch limits are usually from 60° to 70°. If either limit is exceeded, the instrument will tumble or spill and will give incorrect indications until realigned. A number of modern attitude indicators do not tumble.

Every pilot should be able to interpret the banking scale illustrated in Figure 5-13. Most banking scale indicators on the top of the instrument move in the same direction from that in

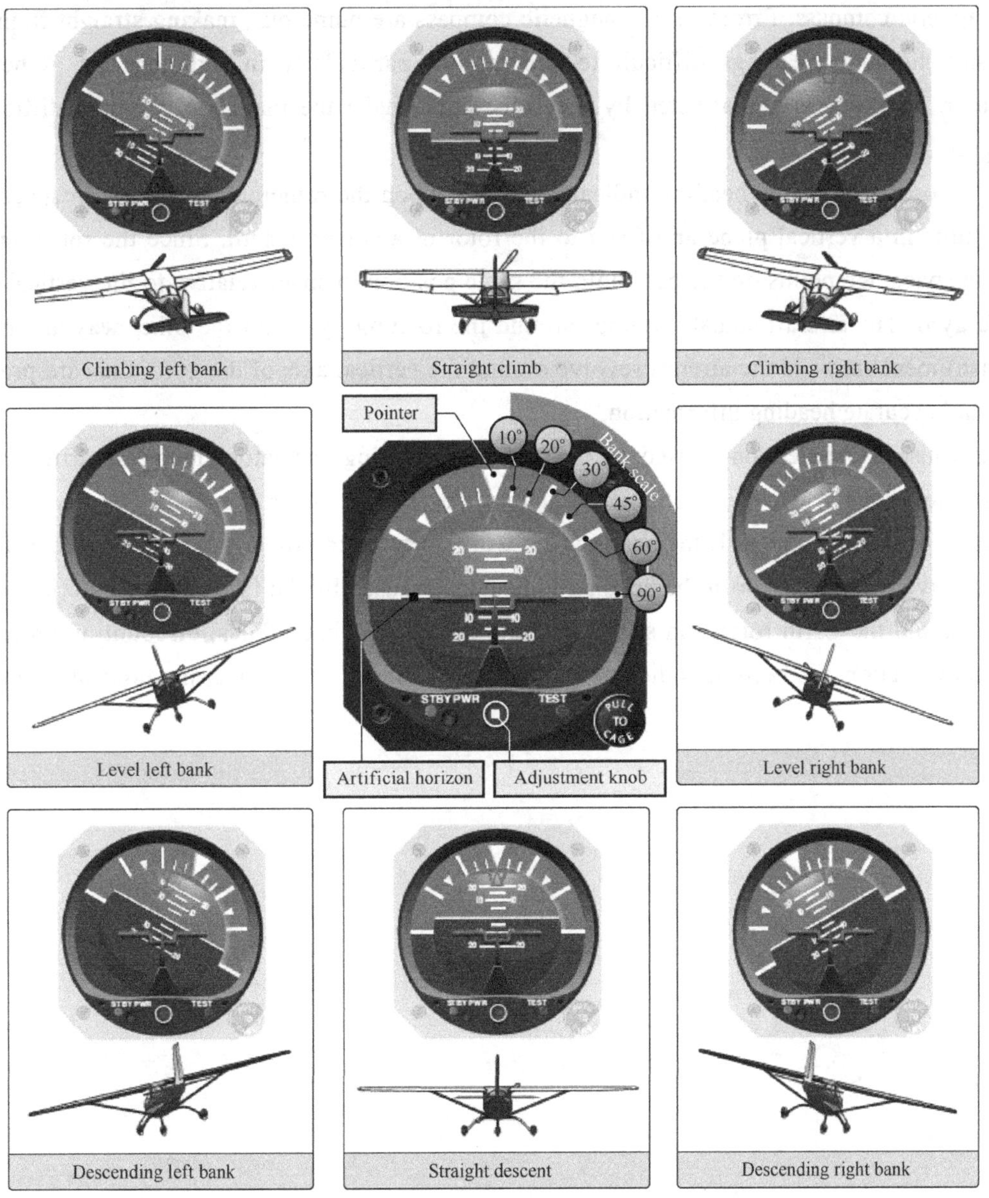

Figure 5-13 Attitude representation by the attitude indicator corresponds to the relation of the aircraft to the real horizon

which the aircraft is actually banked. Some other models move in the opposite direction from that in which the aircraft is actually banked. This may confuse the pilot if the indicator is used to determine the direction of bank. This scale should be used only to control the degree of desired bank. The relationship of the miniature aircraft to the horizon bar should be used for an indication of the direction of bank.

5.2.4 Heading Indicator

The heading indicator is fundamentally a mechanical instrument designed to facilitate the use of the magnetic compass. Errors in the magnetic compass are numerous, making straight flight and precision turns to headings difficult to accomplish, particularly in turbulent air. A heading indicator, however, is not affected by the forces that make the magnetic compass difficult to interpret.

The operation of the heading indicator depends upon the principle of rigidity in space. The rotor turns in a vertical plane and fixed to the rotor of a compass card. Since the rotor remains rigid in space, the points on the card hold the same position in space relative to the vertical plane of the gyro. The aircraft actually rotates around the rotating gyro, not the other way around. As the instrument case and the aircraft revolve around the vertical axis of the gyro, the card provides clear and accurate heading information.

Because of precession caused by friction, the heading indicator creeps or drifts from a heading to which it is set. Among other factors, the amount of drift depends largely upon the condition of the instrument. If the bearings are worn, dirty, or improperly lubricated, the drift may be excessive. Another error in the heading indicator is caused by the fact that the gyro is oriented in space, and the Earth rotates in space at a rate of 15° in 1 hour. Thus, discounting precession caused by friction, the heading indicator may indicate as much as 15° error per hour of operation.

第 5 章　飞 行 仪 表

为了安全驾驶飞机，飞行员必须了解如何解读和操作飞行仪表，也必须能够识别飞行仪表相关的错误和故障信息。本章讨论风速管和相关仪表、真空系统和相关仪表，以及陀螺仪和磁罗盘。只有当飞行员可以清楚地知道每个仪表如何工作并且可以识别仪表的故障时，他才能安全地使用这些仪表并使其充分发挥作用。

5.1　飞行器风速管飞行仪表

风速管是一个综合系统，主要测量空气的静压和由于飞机在空中的运动产生的动压。这些静压、动压给操作空速表(ASI)、高度计和垂直速率表(VSI)提供数据，如图 5-1 所示。

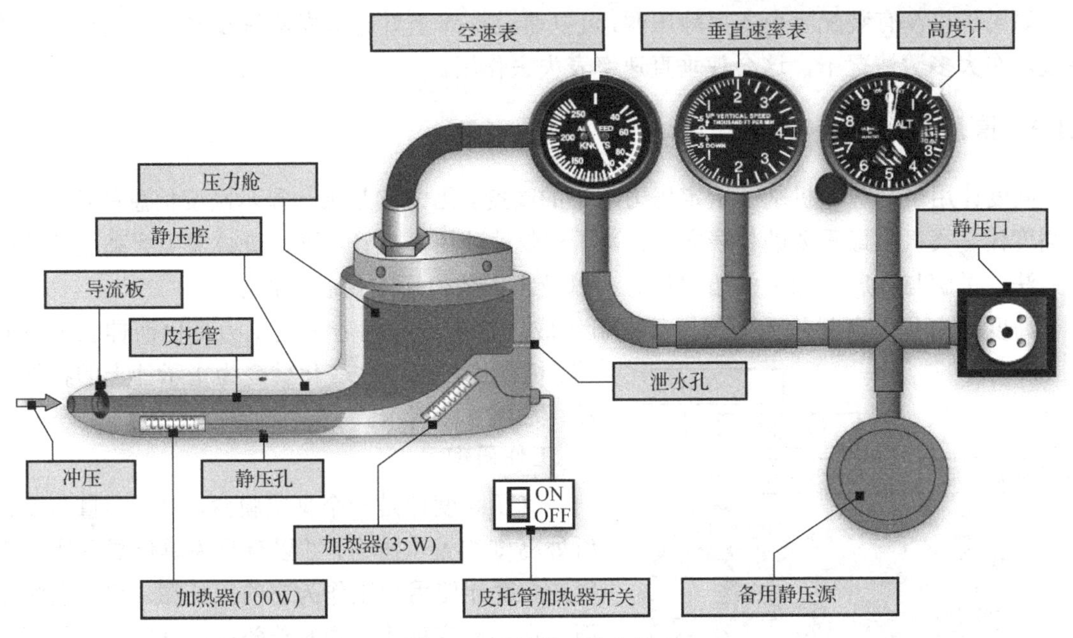

图 5-1　风速管系统和设备

5.1.1　冲压腔和管线

皮托管用来实时测量飞机在空中飞行时的总压。静态压力也称为环境压力，无论飞机处于飞行或者静止状态，静压始终存在。静压来源于局部地区的气压。动压只有当飞机运动的时候才会显示，可以认为它是由于运动产生的压力。空气中风的冲击也会产生动压。无论飞行器以 70 节的速度穿过静止的空气还是飞行器面对以 70 节的速度吹过来的风时，两者产生的动压是相同的。

当风从一个小于 90°的方向吹向机头位置，动压可以被空速表测量出。风以 20 节的速度穿过机翼产生的动压和飞机以 20 节的速度穿过平静空气产生的动压是一样的。皮托管可以同测量大气静压一样实时测量动压。皮托管的前端有一个使总压进入压力舱的小孔。总压等于动压和静压之和。

5.1.2　静压腔和管路

静压腔通过小孔与飞机两侧未受扰动的空气相连。当大气压力增加或者降低时，静压腔中的压力通过连接仪表与静压系统的管路，在仪表内外产生变化并传递到利用静压工作的仪表上。一些飞机在静压开口被堵塞时还提供备用静压源。这个备用静压源通常连通驾驶舱。由于机身周围气流的文丘里管效应，这个备用静压源通常比正常静压空气源的压力低。

当使用备用静压源时，仪表通常会出现以下显示：

(1) 高度计所指示高度略高于实际高度。

(2) 空速表所指示空速大于实际空速。

(3) 当实际高度并没发生变化时，垂直速率表会出现短暂上升然后稳定下来。

飞行员应参考飞机飞行手册或飞行员操作手册来确定使用备用静压源时引入的错误。

如果飞机没有装配一个备用静压源，可以通过打碎垂直速率表密封玻璃使空气进入压力系统。在大多数情况下，这会使垂直速率表失去作用。

5.1.3　高度计

高度计用于测量飞机基于给定压力水平的高度。因为它是唯一显示高度信息的仪表，所以高度计是飞机上最重要的仪表之一。为了有效地使用高度计，飞行员必须了解高度计的操作方法，了解与高度计相关的错误信息，以及它们如何影响仪表的指示。

高度计由几个密封的无液气压计圆盘组成。这些圆盘随着大气压力变化而伸长或者收缩。机械连杆把这些变化转变为指示计上的指针运动，如图 5-2 所示。

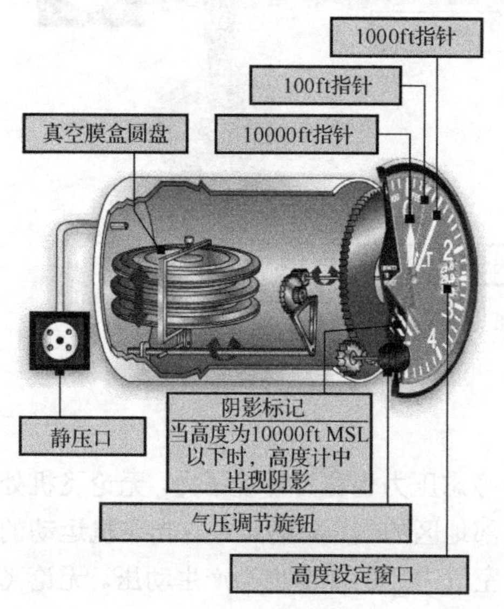

图 5-2　高度计

1. 工作原理

气压高度计是一个真空膜盒，用于测量高度计所处高度的大气压力，并以英尺为单位显示所在高度。高度计使用静压作为它的工作源。空气在海平面比在高空密度大，同时伴随高度增加，大气压力降低。不同高度的压力差异使高度计指示出高度的变化。

不同类型的高度计对高度的显示方式差异很大。有的高度计只有一个指针，而有的高度计有两个或者更多指针。本书只讨论多指针型高度计。典型高度计的表盘刻度按照顺时针方向刻有从 0 到 9 的数字。真空膜盒元件的运动通过齿轮传递到指示高度的三个指针。最短的指针代表万分之一英尺高度，中等长度的指针代表千分之一英尺高度，最长的指针代表百分之一英尺高度。

然而，只有满足以下情况，高度计指示的高度才是正确的，即海平面大气压力为标准的

(29.92in Hg)，海平面大气温度是标准的(15℃或者 59℉)，而且压力和温度以标准速率随高度的升高而降低，通常可以将高度计表面的大气压力刻度设为正确的压力值来完成非标准压力的调整。气压窗口也称为科尔兹曼窗口。只有高度计设置好之后，才会指示正确的高度。

2．非标准压力和温度的影响

如果高度计不将非标准压力进行调整，则可能出现危险情况。例如，如果飞机从高压区飞到低压区的过程中不调节高度计，高度计显示的高度不变，但飞机的实际高度是高于显示高度的。有一句古老的航空格言：“从高到低，注意离地高度。”如果从低压区飞到高压区而不调节高度计，飞机的实际高度会高于指示高度，如图 5-3 所示。

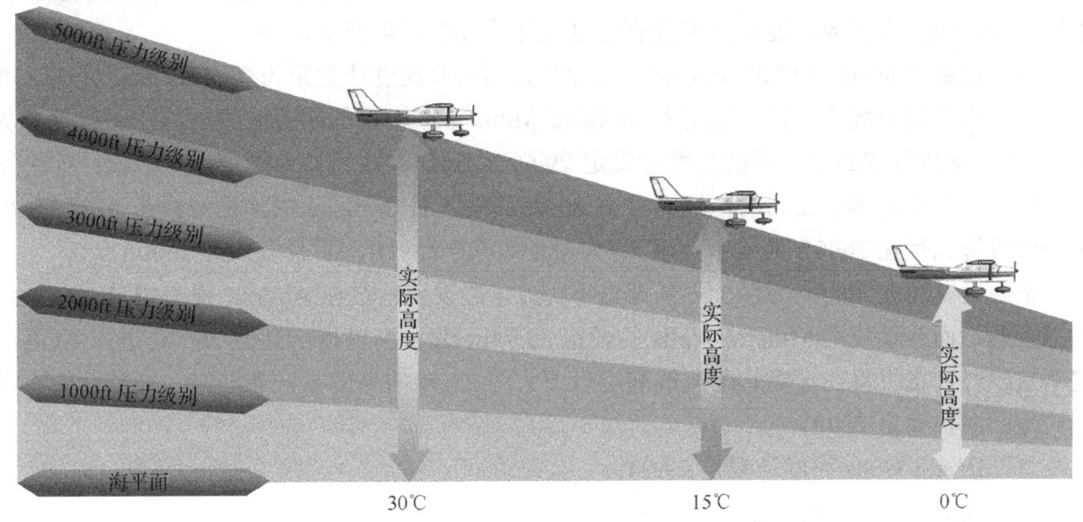

图 5-3 非标准温度对高度计的影响

许多高度计在大气压力超出 31in Hg 高度时无法做出准确的调整。当高度计不能设置为更高的压力设定值时，飞机的实际高度将高于高度计指示。在低气压条件下(低于 28)，如果飞行器的高度计不能设定实际的高度，那么不建议飞行。地形或越障高度是选择巡航高度的一个因素，尤其在多山地区，要考虑到比标准温度更冷的气温会使飞机飞行高度低于高度计指示高度。因此，需要一个较高的指示高度来提供足够的离地高度。我们就应采用“从高到低或者从热到冷，注意离地高度”这个规则。

3．设定高度计

大多数高度计都安装了一个大气压力设定窗口(或科尔兹曼窗口)，用来调节高度计。在仪表的底部位置有一个调节旋钮。

为了根据大气压力的变化调节高度计，高度计设定窗口的压力数值是以英寸汞柱或毫巴为单位，设定窗口的压力数值要与高度计的设定值匹配。高度计所设定的参数定义为气象站压力减去海平面压力。但是，高度计所设定的参数仅在气象站附近是准确的。因此，高度计所设定的参数必须随着飞行进程从一个气象站调整到下一个气象站。

很多飞行员确信当前高度计设定能够补偿所有高度上大气压力的无规律变化，但这不并总是正确的。地面站附近的高度计设定广播是修正平均海平面的气象站压力。它不能解决高飞行高度时气压的不规则性问题，特别是非标准温度的影响。然而，如果在一个给定区域的每个飞行员使用相同的高度计设定，那么每个高度计受温度和压力变化误差的影响应该是相同的，在飞机之间维持预期的垂直间隔就成为可能。飞机处于山群地形时，特定的大气状况可能导致高度计指示比实际高度高出1000ft，或者更多。因此，考虑到可能的高度计误差以及和强风所带来的向下气流，飞行员需要留有较大的高度余量。

为说明如何使用高度计设定系统，我们跟随一趟航班，从德克萨斯州的达拉斯 Love Field 机场经由 Mineral Wells 飞到德克萨斯州的 Abilene Municipal 机场。在 Love Field 机场起飞前，飞行员从控制塔台或者自动终端信息服务(ATIS)收到当前高度计设定为29.85in Hg，并在高度计设定窗口中设定这个值。此时，飞行员应该把高度计指示值和已知的机场高度487ft 进行比较。因为大多数高度计如果没有经过很好的校正，会产生误差。

当飞行经过 Mineral Wells 时，假设飞行员收到当前高度计设定为29.94in Hg，并在高度计窗口中设定这个数值。在飞机进入 Abilene Municipal 机场的起落航线之前，飞行员会从 Abilene 控制塔台收到一个新的高度计设定29.69in Hg，并在窗口中设定这个数值。如果飞行员预期飞行的起落航线大约在地面以上800ft 高度，且 Abilene 的地面海拔是1791ft，那么应该维持飞行在2600ft 的指示高度上(1791ft+800ft=2591ft，四舍五入为2600ft)。

正确设定高度计的重要性一再强调也不为过。假设飞行员在 Abilene 没有调节高度计到当地参数，而继续使用 Mineral Wells 的设定29.94in Hg。当以指示高度2600ft 进入 Abilene 起落航线时，飞机将在正确起落航线高度以下大约250ft 的高度。在着陆时，高度计的指示值比地面海拔高出250ft。

(1) Mineral Wells 高度计设定29.94。

(2) Abilene 高度计设定29.69。

(3) 差值0.25。

(1in 压力大约相当于1000ft 高度变化，0.25×1000ft=250ft。)

4. 高度计的运行

高度计的指针可以通过两种方法来调节。一种是气压的变化，另一种是调节大气压力刻度。当飞机爬升或者下降时，高度计容器中的压力变化使真空膜盒膨胀或者收缩。这个运动经过机械连杆传递使指针旋转。

压力降低导致高度计指示高度的增加，压力增加导致高度计指示高度降低。飞行员必须知道飞机的高度对飞行员是至关重要的。飞行员必须确保飞机飞行在足够的高度，以避开最高的地形或者沿预期航线的障碍物。当能见度受限时，准确的高度信息尤为重要。为避开障碍物，飞行员必须随时了解飞机的高度和周围地形的海拔变化。为降低飞行器在空中发生碰撞的可能性，飞行员与空中交通规则要求的高度保持一致是很重要的。

5. 飞行高度类型

飞行高度是指高于某一个参考点或参考平面的垂直距离。根据测量参考平面的不同有多种类型的高度，每一种都可以用于特定的情况。飞行员需要掌握五种高度。

(1) 指示高度：当高度计设定为当前高度计设定时，飞行员可以直接从表(未校正的)上读出的高度数值。

(2) 真实高度：飞机距离海平面的垂直距离，即实际高度。它通常表示为距离平均海平面(MSL)之上的英尺数。航图上机场、地表和障碍物的高度是真实高度。

(3) 绝对高度：飞机在地表之上的垂直距离，或者距离地面(AGL)的垂直距离。

(4) 压力高度：当高度计设定窗口(大气压力数值)调节到29.92in Hg时的指示高度。这是标准基准面之上的高度，该基准面是一个气压(当环境温度为15℃时)等于29.92in Hg的理论平面。压力高度用于计算密度高度、真实高度、真空速和其他性能数据。

(5) 密度高度：根据标准温度的变化而校正的压力高度。当处于标准环境条件时，压力高度和密度高度相同。如果温度高于标准条件，那么密度高度高于压力高度。如果温度低于标准条件，那么密度高度低于压力高度。密度高度是一个重要的高度，因为它和飞机性能直接相关。

举一个例子，一个机场地面距离平均海平面5048ft，标准温度为5℃。在这种条件下，压力高度和密度高度相同，都是5048ft。如果温度变为30℃，密度高度就增加到7855ft。这就意味着飞机在起飞时将表现为类似在标准温7855ft的机场起飞的状态。相反，–25℃的温度将使密度高度变为1232ft。飞机在这种条件下将会表现出更好的性能。

6. 仪表检查

为确定高度计的状况，常常需要把大气压力数值设定为本地自动化飞行服务站(AFSS)或任何其他可信来源(如ATIS、AWOS或ASOS等)传来的高度计数据参数。高度计指针应处于机场实测高度的数值位置。如果指示高度和测量海拔偏差大于75ft，这个仪表就应该交付认证的仪表维修站来校正。

5.1.4　垂直速率表

垂直速率表(VSI)也称为垂直速度表(VVI)，它显示飞机是在爬升、下降还是在水平飞行。爬升或者下降速率以英尺每分钟为单位显示。如果经过正确的校正，在水平飞行时垂直速率表显示读数为0，如图5-4所示。

垂直速率表仅靠静压工作，是压差仪表。它包含一个通过连杆和齿轮连接到密封盒内指针的隔膜。隔膜的内部直接与全静压系统的静压管相连。在仪表盒子里面的隔膜外部区域也将要通过一个受限制的孔(校正的漏气口)连接到静压管。

在现有的大气压力下，隔膜和盒子都接受来自静压管的空气。当壳体通过计量泄漏接受静压时，将不再约束空气的进出，即失效。当飞机在地面或者水平飞行时，隔膜和仪表盒子内部的压力仍然相同，指针位于0位置。当飞机爬升或者下降时，隔膜内部的压力立即改变，但是由于仪表的计量存在时延，短时间内盒子压力仍然保持较高或者较低，导致隔膜收缩或者膨胀。这产生了压力差，表现在仪表指针上就是指示为爬升或者下降。当压力差稳定在一定速率后，指针将代表高度变化的速度。

垂直速率表显示两类信息：

(1) 实时显示飞机爬升的速率是增加或是降低。

(2) 速率信息显示飞行高度变化是否稳定。

从爬升率的开始变化到垂直速率表准确显示改变后的爬升率，这段时间被称为延迟(或者叫间隔)。不熟练的控制技术和紊流会延长间隔时间，导致无规律且不稳定的速率指示。一些飞机装配了瞬时垂直速率表(IVSI)。它结合加速计来补偿典型垂直速率表中的延迟，如图5-5所示。

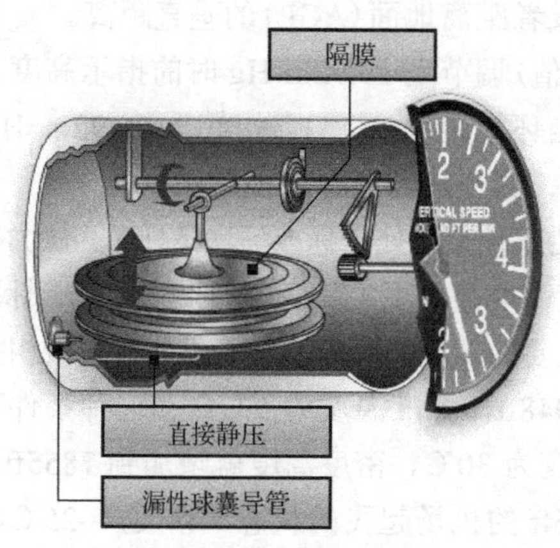

图 5-4　垂直速率表

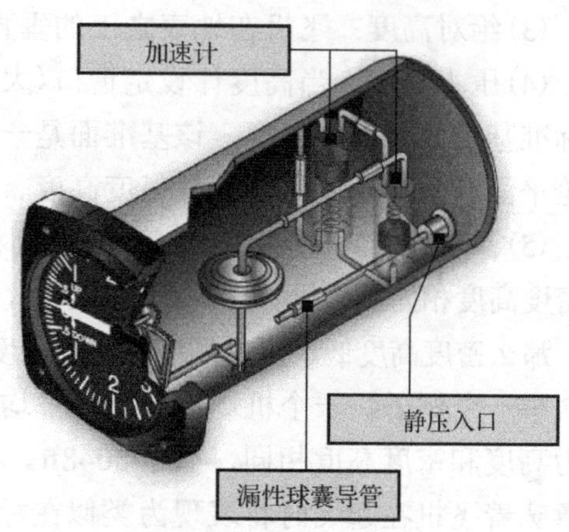

图 5-5　用瞬时垂直速率表和加速计来显示垂直速度的变化

5.1.5　空速表

空速表是一个灵敏的差压表，用来测量并及时显示皮托(冲/动压)和静压之间的差值，如图 5-6 所示。在无风情况下，飞机停放在地面时，这两个压力相同。当飞机在空气中飞行时，皮托管上的压力大于静压管中的压力。这个压力差被空速指针表示在仪表盘面上，它以英里/小时(1mile=1.609344km)、节(海里/小时)(1n mile=1.852km)为刻度单位。

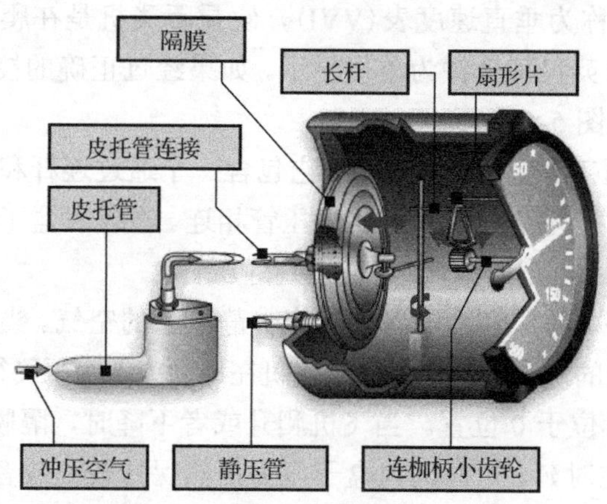

图 5-6　空速表

飞行员要对以下速度非常熟悉.

(1)指示空速(IAS)：在没有根据大气密度变化、安装误差或仪表误差来进行校正的情况下，直接从空速表上获得的仪表读数。制造商使用这个空速作为确定飞机性能的基准。在飞机飞行手册或者飞行员操作手册中列出的起飞、着陆和失速速度都是指示空速，一般不随海拔或温度而变化。

(2)标定空速(CAS)：校正安装误差和仪表误差之后的指示定速。尽管制造商将空速误差控制在最低限度，但消除空速运行范围内的所有误差是不可能的。在给定空速和襟翼参数的情况下，安装误差和仪表误差可能有几节。这个误差通常在低空速时最大，而在巡航和较高空速范围内，指示空速和标定空速大致相同。请参考空速校正图标来纠正可能的空速误差。

(3) 真空速(TAS)：针对高度和非标准温度进行校正后的标定空速。因为空气密度随高度增加而降低，飞机在更高的高度上必须飞得更快才能在皮托冲压和静压之间产生相同的压力差。因此，对于一个给定的标定空速，真空速随海拔增加而增加；或者对于一个给定的真空速，标定空速随海拔增加而降低。飞行员可以用两种方法获得真空速。最准确的方法是使用领航计算尺。对于这种方法，标定速度是通过使用领航计算尺上的空速修正数值根据温度和压力变化来修正的。也可以使用高精度的电子飞行计算器。只需要输入标定空速、压力高度和温度，计算机就会计算真空速。第二种方法是“拇指规则”，该方法可以提供近似的真空速。只需每 1000ft 高度标定空速增加 2%即可。真空速是写入飞行计划的速度，用于提交飞行计划。

(4) 地面速度(GS)：地面速度指飞机相对于地面的实际速度。它是根据风速来调整的真空速(注：根据风速做出修正后的真空速，这个速度考虑地面作为速度参照物)。地面速度在迎风时减小，顺风时增加。

1. 空速表标记

1945 年以后制造的，重量不超过 12500 磅且被美国联邦航空局(FAA)认证的飞机，要求其空速表按照标准彩色编码标记系统来印标。这个彩色编码标记系统使飞行员看一眼就知道对飞机安全飞行极为重要的空速限制。例如，在执行任务期间，注意到空速指针处于黄色弧线内，并迅速接近红色线，直接反应是应该降低空速。如图 5-7 所示，单发小型飞机上的空速表包含下列标准彩色编码标记。

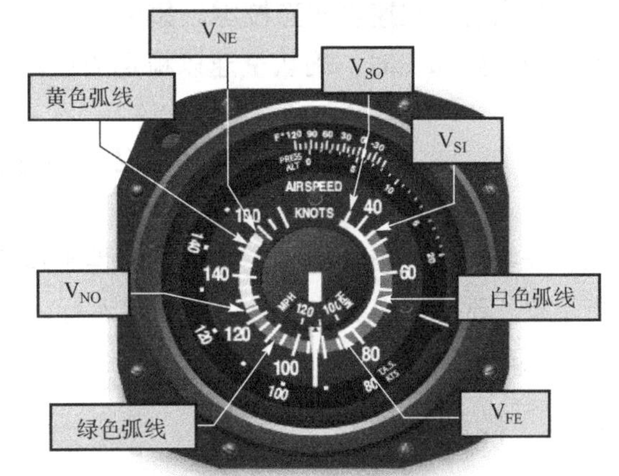

图 5-7　空速表标记

(1) 白色弧线：通常指襟翼运行范围，它的下限表示完全襟翼失速速度，上限表示最大襟翼速度。前进和着陆速度通常在白色弧线速度范围内。

(2) 白色弧线的下限(V_{SO})：飞机着陆时的失速速度或者最小稳定飞行速度。在小型飞行机上，这是着陆配置(起落架和襟翼都放下)中最大着陆重量下的停车失速速度。

(3) 白色弧线的上限(V_{FE})：襟翼伸展时的最大速度。

(4) 绿色弧线：飞机的正常运行速度范围。大多数飞行处于这个速度范围。

(5) 绿色弧线的下限(V_{SI})：特定配置下获得的失速速度或者最小稳定飞行速度。对于大多数飞机来说，这是最大起飞重量下低阻配置(若襟翼可伸缩，则起落架收起，襟翼也收起)的停车失速速度。

(6) 绿色弧线上限(V_{NO})：最大结构巡航速度(注：超过这个速度可能引起飞机部分结构应力过载)。只有在稳定气流中飞行时可以超越该速度，其他情况不要超过这个速度。

(7) 黄色弧线：警告范围。只有在稳定气流中飞行时才可以处于这个速度范围。

(8) 红线(V_{NE})：决不超过的速度。禁止在这个速度以上运行，因为它可能导致结构损坏或结构故障。

2. 其他空速限制

一些重要的空速限制没有标记在空速表的表盘上，但可以在标牌和飞机飞行手册或飞行员操作手册上找到。这些空速包括如下。

(1) 设计机动速度(V_A)：不导致结构损坏的结构设计极限载荷的最大速度(遭遇阵风或控制舵面的满舵偏转)。参考这个速度时需要考虑重量。例如，当飞机有较重的载荷时 V_A 可能是 100 节，但是载荷轻时只有 90 节。

(2) 起落架操作速度(V_{LO})：对于装配有收放起落架的飞机而言，该速度就是伸出或者收缩起落架的最大空速。

(3) 起落架伸出速度(V_{LE})：飞机在起落架伸出后可安全飞行的最大空速。

(4) 最佳爬升角速度(V_X)：飞机能够在给定的距离内获得最大高度的空速。这个速度在短场起飞飞越障碍物时使用。

(5) 最佳爬升率速度(V_Y)：飞机以该空速能够在给定时间内获得最大高度。

(6) 单发失效时的最佳爬升率速度(V_{YSE})：在轻型双发飞机有一个发动机失效时，在给定时间内能够获得最大高度的空速。它在空速表中通常标注为蓝线。

(7) 最小控制速度(V_{MC})：这是轻型双发飞机在单发失效时，另一个发动机在起飞功率状态时，飞行员可以控制飞机所需的最小空速。

3. 仪表检查

起飞前，空速表应当校零。如果有一股直接吹向皮托管的强风，空速表的读数可能比 0 大。当开始起飞时，确认空速以稳定的速率逐步递增。

5.2　陀螺飞行仪表

一些飞行仪表利用了陀螺仪的特性来运行，最常见的有转弯协调仪、航向指示仪和姿态指示仪。要想知道这些飞行仪表如何运行，需要掌握仪表动力系统、陀螺原理和每个仪表的工作原理。

5.2.1　陀螺原理

任何旋转物体都具有陀螺的特性。利用陀螺的特性来设计和安装的轮子或转子称为陀螺仪。仪表陀螺的两个重要设计特性是较大的质量或密度，以及高速旋转时的低摩擦力。

一般来说，有两种通用类型的装配结构，具体使用哪种类型取决于要利用陀螺仪的哪个特性。自由安装的陀螺仪能够自由地绕它的重心以任意方向旋转。这样的陀螺转子有三个自由度平面。轮子或转子在任何一个与支架相关的平面内自由旋转，陀螺转子在静止时也是平衡的，它会保持在被放置的位置。受限或半刚性安装的陀螺仪是指安装在基座上并使其中一个自由面固定在基座上的陀螺仪。

陀螺效应固有特性：空间刚度和进动性等。

1. 空间刚度

空间刚度是指陀螺仪在旋转时将会保持在一平面不变，如图 5-8 所示。

把轮子或陀螺仪安装在一组万向环上，陀螺仪能够在任何方向自由旋转。因此即使万向环是倾斜的、螺旋的或移动的，陀螺仪还会保持在它最初所旋转的平面内。

2．进动性

进动性是陀螺对偏转力的反馈从而形成的倾斜或者旋转。对偏转力的反作用不是发生在它所施加的点上，而是发生在旋转90°以后的点上。这个原理使陀螺能够通过检测方向变化产生的压力大小来确定旋转的速度。陀螺进动的速度和旋转速度成反比，和偏转力大小成正比，如图5-9所示。陀螺的进动性在一些仪表中也会造成微小的误差。

图5-8 不考虑基座的位置，旋转坐标轴指向固定的方向，陀螺趋向于空间保持刚性

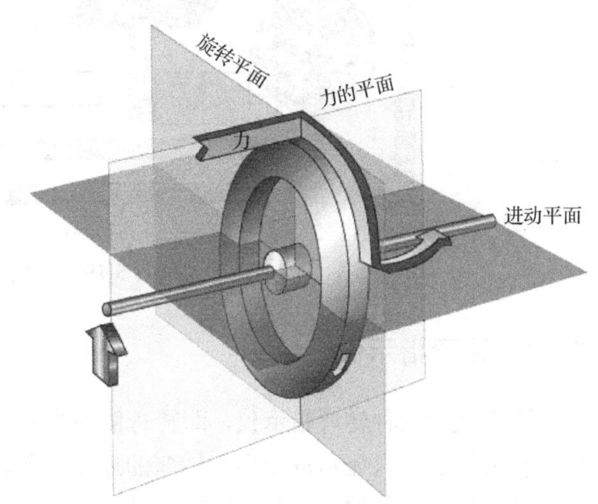

图5-9 施加偏转力使陀螺产生进动

3．动力源

在某些飞机上，所有陀螺仪都是真空、压力或电动的；而其他飞机，真空系统和压力系统为航向指示仪和姿态指示仪提供动力，而电气系统为转弯协调仪提供动力。大多数飞机至少有两个动力源，以确保一个动力源失效时还有一个动力源。真空或者压力系统通过吸入一股高速气流来冲击转子环，通过高速旋转转子来旋转陀螺，其原理类似水车或者涡轮机的运行。仪表运行所需的真空或者压力大小是变化的，但是通常位于4.5～5.5in Hg范围内。

陀螺仪的真空源之一就是安装在发动机附件箱上的环形发动机驱动泵。不同飞机的发动机驱动泵容量不同，它取决于陀螺仪的多少，如图5-10所示。

典型的真空系统由发动机驱动的真空泵、减压阀、空气过滤器、量表和完成连接所需的管路组成。量表安装在飞机的仪表面板内，并指示出系统内部的压力大小(真空用英寸汞柱来标示，通常低于周围环境压力)。如图5-10所示，空气被发动机驱动的真空泵抽进真空系统。首先经过一个过滤器，它能防止外面的物体进入真空或压力系统。然后，空气流经姿态指示仪和航向指示仪，在这里它使陀螺仪旋转。使用减压阀可以防止真空压力或者抽气机超过指定的限制。之后，空气被排出系统或者用在其他系统内，如为除冰气源系统充气。

飞行期间监测真空压力很重要，因为吸气压力低的时候姿态指示仪和航向指示仪可能不能提供准确的信息。真空、吸气或量表会在正常刻度范围内给予标记。一些飞机装配了警示灯，当真空压力下降到低于标准值的时候警示灯就会发亮。

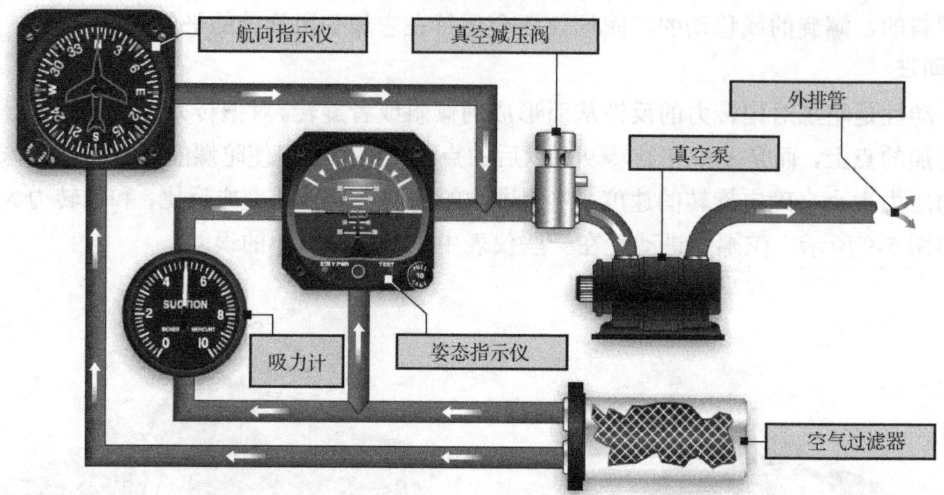

图 5-10　典型真空系统

5.2.2　转弯指示仪

飞机使用两种转弯指示仪，即转弯侧滑指示仪和转弯协调仪。受陀螺仪安装方式的影响，转弯侧滑指示仪只显示每秒指示转弯的速度。转弯协调仪上的陀螺仪与安装面有一定的倾角，或者说是倾斜安装的，所以开始时它显示滚转速度。一旦滚转稳定后，它就显示转弯的速度。两个仪表都显示转弯方向和质量(转弯协调性)，也可以用作姿态指示仪失效时的备用信息来源。协调性是通过使用倾角计获得的，它由充满液体的弯管组成，其中有一个小球，如图 5-11 所示。

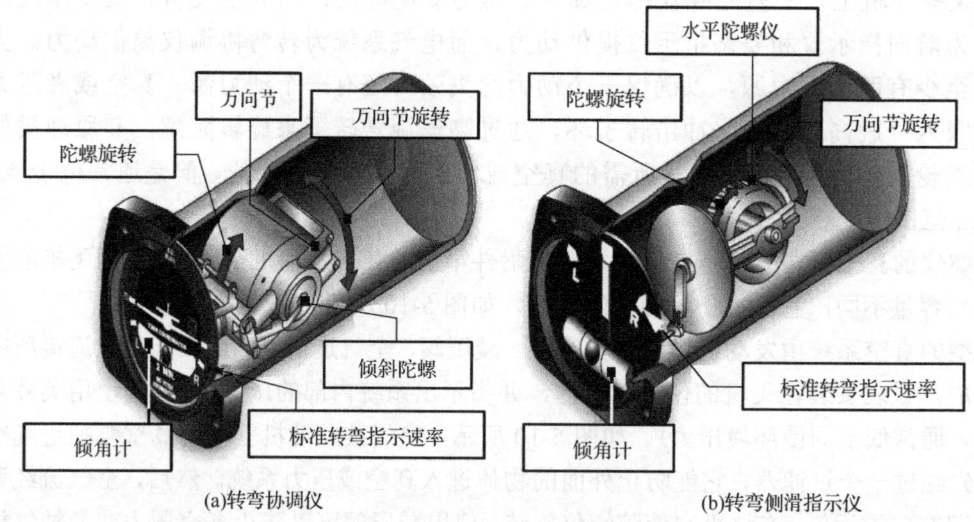

图 5-11　转弯指示仪依赖于操作所产生的可控进动

1. 转弯侧滑指示仪

转弯侧滑指示仪中的陀螺仪在竖直平面内旋转，对应于飞机的纵轴。单极万向节把陀螺仪限制在其中倾斜的平面内，而弹簧为其回到中心提供恢复力。由陀螺仪的进动性水平方向的偏转力使得陀螺仪从飞行员座位看去是向左或者向右倾斜的。转弯侧滑指示仪使用转弯指针来显示转弯的方向和速度。

2．转弯协调仪

转弯协调仪中的万向节是倾斜的。它的陀螺仪可以检测滚转速度和转弯速度。由于在教练机上更多的使用转弯协调，所以这里集中讨论这个仪表。在转弯侧滚或者转弯退出侧滚时，小型飞机会向飞机侧滚方向倾斜。小型飞机在侧滚速度较快的情况下，倾斜会更明显。

转弯协调仪通过使小型飞机的机翼和转弯指针相互协调与配合来建立和维持标准速率转弯。

3．倾角计

倾角计用于表示飞机的偏航，它所描述的是飞机机头从一侧到另一侧的运动。在协调转弯和平直飞行时，重力使倾角计的小球保持在弯管的参考线中央。协调转弯飞行是通过保持小球居中而维持的。如果小球没有居中，可以通过控制方向舵来使其居中。

要使小球居中，就得在球偏移的一侧施加方向舵压力。使用简单的规则"脚踏球上"来记住应该踩哪边的脚舵。(注：小球在右边，就踩右边脚舵来居中，否则踩左边脚舵。)如果副翼和方向舵在转弯时是协调的，小球就会保持在弯管的中间。如果空气动力不平衡，小球就会离开弯管的中线。内滑时，转弯速度对于倾斜角来说过慢，小球就会向转弯的内侧移动。外滑时，转弯速度对这个倾斜角来说过快，小球向转弯的外侧移动。为纠正这种状态，改进转弯的质量，记住"脚踏球上"。改变倾斜角可以使飞行器从外滑或内滑中恢复协调飞行。要纠正内滑，可以降低倾斜角或者增加转弯速度。要纠正外滑，可以增加倾斜角或者降低转弯速度。

4．仪表检查

飞行前，要检查并确保倾角计充满液体且没有气泡，且滚珠处于它的最低点。当转弯发生侧滑时，转弯协调仪应该指示正确的转弯方向，而小球向与转弯方向相反的方向移动。

5.2.3　姿态指示仪

姿态指示仪以模拟飞机与地平线的相对位置来显示飞机的姿态情况。模拟飞机与地平线的相对位置关系与真实飞机与实际地平线的关系是相同的。仪表实时显示飞机姿态的变化，哪怕是微小的变化。

姿态指示仪中的陀螺仪安装在水平平面上，它的运行取决于空间刚度。仪表上的地平线线条表示真实地平线。这个地平线被固定在陀螺仪上，并且保持在水平平面内，当飞机绕它的横轴或者纵轴俯仰或者倾斜时，它能够指示飞机相对于真实地平线的姿态。

飞行员可以通过一个调节旋钮来调节微型飞机对应于地平线的上下位置，以配合飞行员的视线。通常，微型飞机采用这种调节方式来确保在平直飞行时机翼与地平线平行(图 5-12)。

仪表盘的俯仰与倾斜限制取决于仪表的品牌和型号。倾斜平面的限制通常为 100°～110°，俯仰限制通常为 60°～70°。如果超过任何一个限制，姿态指示表将会混乱，直到重新稳定才会正确显示读数。不过，目前的现代姿态指示仪不会出现混乱情况。

倾斜指数
陀螺旋转
侧滚陀螺
参考水平线
陀螺
俯仰陀螺

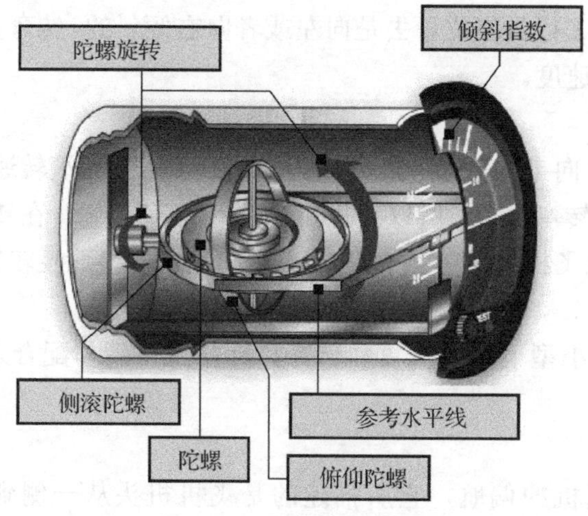

图 5-12　姿态指示仪

每个飞行员都应该能解读如图 5-13 中所示的倾斜刻度。大多数仪表顶部倾斜刻度指示仪和飞机实际倾斜的方向同向运动，但也有一些移动方向和飞机实际倾斜方向相反运动。上述情况往往会使飞行员迷惑。这个刻度只能用于控制期望飞机的倾斜角度。仪表盘上的微型飞机与地平线的关系通常为飞行员指示飞机倾斜方向。

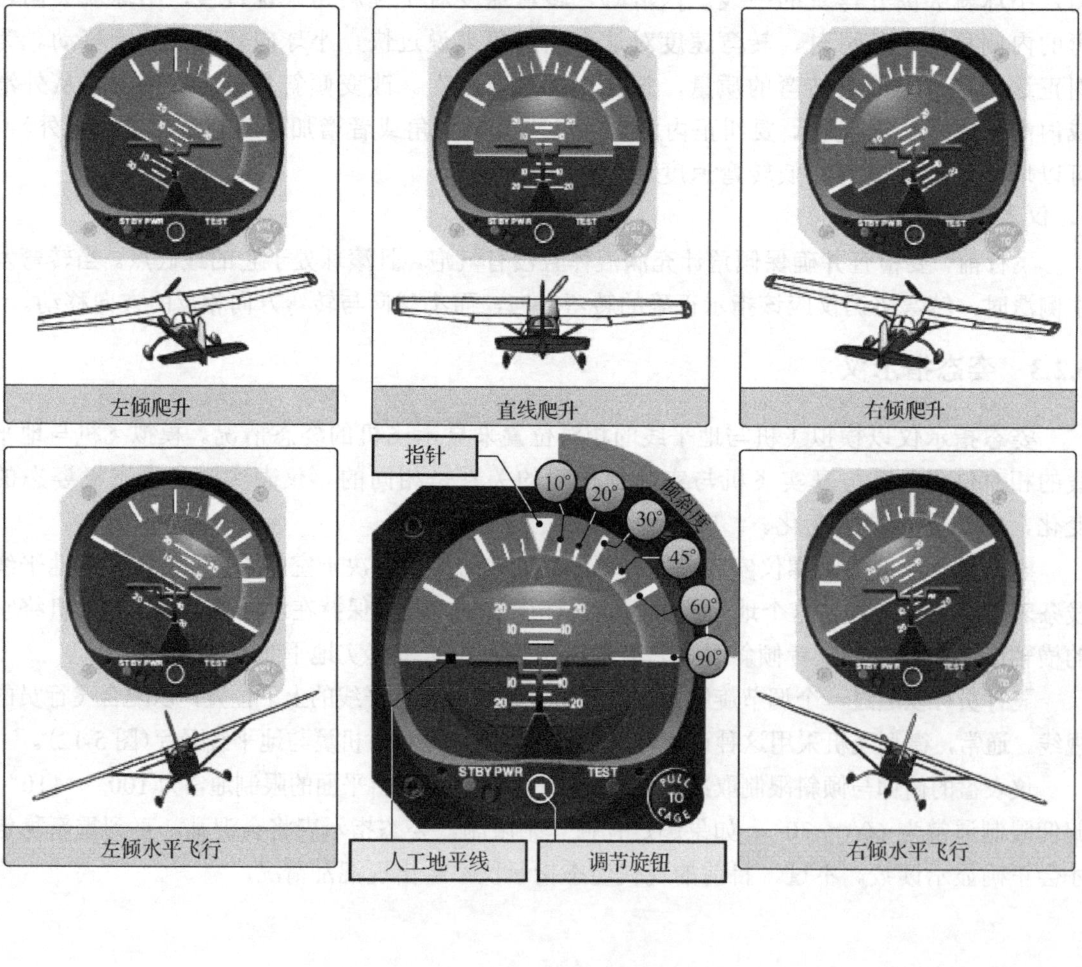

左倾爬升
直线爬升
右倾爬升
指针
10°
20°
30°
45°
60°
90°
倾斜度
STBY PWR
TEST
PULL TO CAGE
左倾水平飞行
人工地平线
调节旋钮
右倾水平飞行

左倾下降

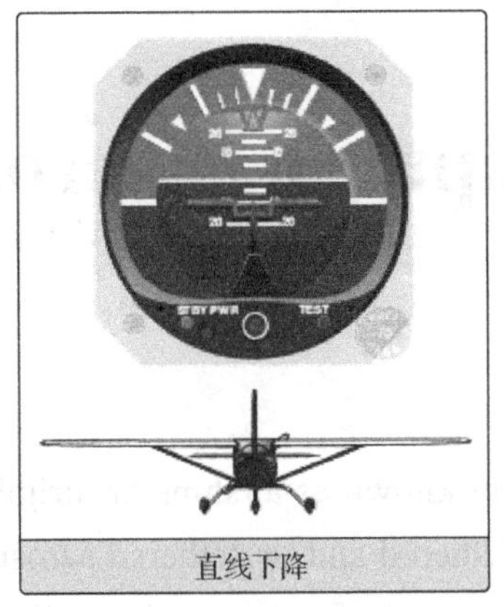

直线下降

右倾下降

图 5-13 通过姿态指示仪的姿态显示来了解飞机的真实运行状态

5.2.4 航向指示仪

航向指示仪是一个基本的机械仪表，它配合磁罗盘来使用更方便。磁罗盘中误差很多，这使直线飞行和精确转弯到特定航向难以完成，尤其是飞机处于紊流中时，误差更多。外力会使磁罗盘难以准确指示，然而航向指示仪并不会因此受到影响。

航向指示仪的运行依据是空间刚性原理。转子在一个垂直平面上转动，并固定在一个罗经刻度盘的回转轴上。因为转子保持空间内的刚性，刻度盘上的点在空间内保持相对于垂直平面的相同位置。飞机实际上是围绕着旋转的陀螺仪同圈旋转而非其他方式旋转。只有当仪表盒子和飞机绕陀螺垂直轴旋转时，刻度盘才能为飞行员提供清晰、准确的航向信息。

摩擦力引起的进动会导致航向指示仪从一个航向缓慢移动或漂移到设定的航向。在其他因素中，漂移量很大程度上受仪表状态影响。如果轴承磨损、脏或润滑不当，则飞机在转弯时会过量。航向指示仪中的另一个误差是由于陀螺仪的空间导向所导致。由于地球在太空的旋转速率是 15°/h，于是不计摩擦引起的进动，航向指示仪可能会指示每小时运行约 15° 的误差。

Chapter 6 Aerostat

6.1 Introduction

Lighter-than-air vehicles, popularly known as airships or dirigibles, started with hot air balloons and evolved to lifting gas filled, tethered and un-tethered aerostats, airships, and novel buoyancy air vehicles in step with the advancement of new materials and technologies.

The aerostat is an aircraft that is internally filled with gas lighter than air, relying on or primarily depending on the static buoyancy of the air to achieve liftoff. The aerostats are mainly divided into two categories: balloons and airships.

In the mid-1990's, the first aerostat concept made use of a commercial off-the-shelf coated material that met the flexibility and helium retention requirements. The aerostat was sized with this material in mind and such that it could be handled during inflation, deflation and flight with two to four people. The design utilized the same material for the hull, fins, ballonet and reinforcements. The fins were filled with helium from the hull.

The aerostat rigging and hardware was simple. The payload was suspended within the flying (confluence) lines so that a variation in the payload weight did not affect the aerostat's pitch angle. A continuously running fan kept the ballonet filled at its stall pressure. A quick-disconnect fitting for helium inflation was installed in a plate bolted to a ring set; the plate was removed for deflation.

The mooring system was a tether winch mounted on the back of a truck along with a generator and slip rings to provide power and signal to and from the aerostat. When the aerostat was not flying, this compact design required the aerostat to be moored away from the truck on a cable bridle. The mooring system equipment was moved to a trailer so that the truck could be driven away during operations while also providing greater storage capability. The aerostat was modified with different fin shapes and sizes to improve the flight characteristics while the material was changed to improve its ruggedness and the rigging was improved to increase its reliability.

6.2 Balloons

6.2.1 Classification of Balloons

Balloons can generally be divided into tethered balloons, divided into two types, the conventional tethered balloon and the manned tethered balloon respectively, and free balloons that include hot air balloons and air balloons.

The conventional tethered balloon mainly relies on the gas lighter than air that is filled into the inside of the airbag to generate buoyancy, overcomes its own weight and floats in the air, and tied to the ground position or the traction tether, using the cable. The ground winch system, controlling the floating height of conventional tethered balloons, can also achieve the rise and fall of theirs. At present, conventional tethered balloons mostly use one of the safe inert gas called helium as their buoyancy-generating medium. The system of conventional tethered balloons is mainly composed of an air ball, a tethered cable assembly, an anchoring facility, and a ground comprehensive support facility. Tethered balloons can also be classified into mobile, positional or ship-type tethered balloons, depending on their characteristics of use (Figure 6-1).

The principle of a manned tethered balloon is similar to that of a conventional tethered balloon. Unlike conventional tethered balloons, which are designed for a specific mission, manned tethered balloons are mostly entertaining, so they are not required for flight height, but have higher requirements for flight performance. Manned tethered balloons are generally composed of balloons, ground equipment and steering equipment. While descending, they mainly rely on the ground winch to drive the balloon back to the landing platform on the ground (Figure 6-2).

Figure 6-1 The first practical car-mounted tethered balloon of China

Figure 6-2 Manned tethered balloon

6.2.2 Tethered Balloon Technology

During a natural disaster, only space communication technologies can mitigate the impact when the infrastructure is destroyed. Therefore, space technologies play a vital role in recovering from all types of catastrophes. They are used for collecting data needed to protect humans and reduce economic losses. In the immediate aftermath of a disaster, the satellite is one of the reliable solutions of communication. However, delay and launching costs are the reasons for the weakness of satellite communication. Therefore, the use of an aerial platform is the best solution, taking into consideration the merits of satellite and terrestrial wireless communication systems. The categories of the aerial platform include high-altitude platforms (HAP), medium-altitude platforms(MAP), and low-altitude platforms (LAP).

Tethered balloon technology belongs to the LAP family. It operates at an altitude of 200—440m above the ground. The significant advantages of tethered balloon are the low cost of deployment, low propagation delay, rapid deployment, fixed station, and especially its use in case of disasters. The implementation of the tethered balloon for emergency situations is vital for natural and human-induced disasters. These advantages of tethered balloon make it a more attractive concept for emergency communication through rapid deployment and for users to operate their existing mobile handsets in disaster regions. Therefore, it represents the best solution for disaster recovery, supporting the relief and rescue teams to perform their tasks effectively and efficiently. The payload of the tethered balloon includes global system for mobile communication (GSM), code division multiple access (CDMA), ad hoc, long term evolution (LTE), universal mobile telecommunication system (UMTS), and so on, as shown in Figure 6-3.

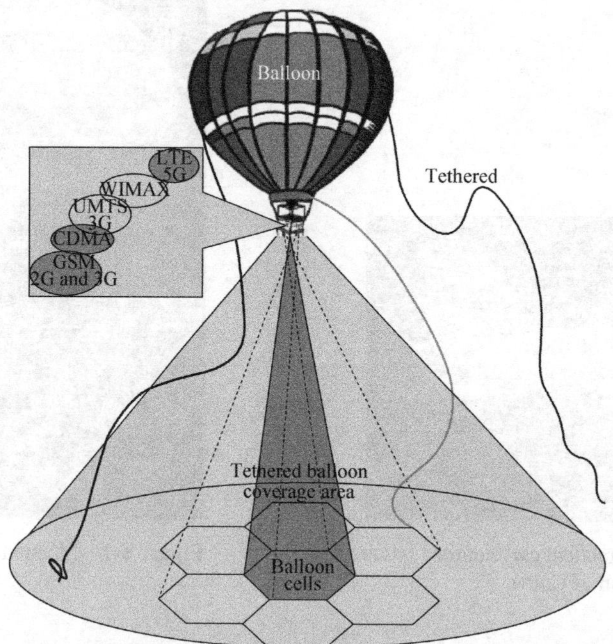

Figure 6-3　Payload of tethered balloon

Saeed Hamood Alsamhi et al. discussed the tethered balloon technology used in design solutions for rescue and relief team emergency communication services. A tethered balloon is a balloon that is tethered to the ground by using ropes. The length of the tethered balloon is limited to the maximum balloon altitude. To maintain the stability of the balloon against the wind, rose ropes are used. The number of ropes depends on the wind speed forecast and altitude target.

The operation of a tethered balloon is at an altitude of 200–440m. The coverage provided by a tethered balloon is 5.5km radius or 72km^2 from an altitude of the balloon of 440m. Coverage can be extended by increasing the altitude of the balloon, as shown in Figure 6-4.

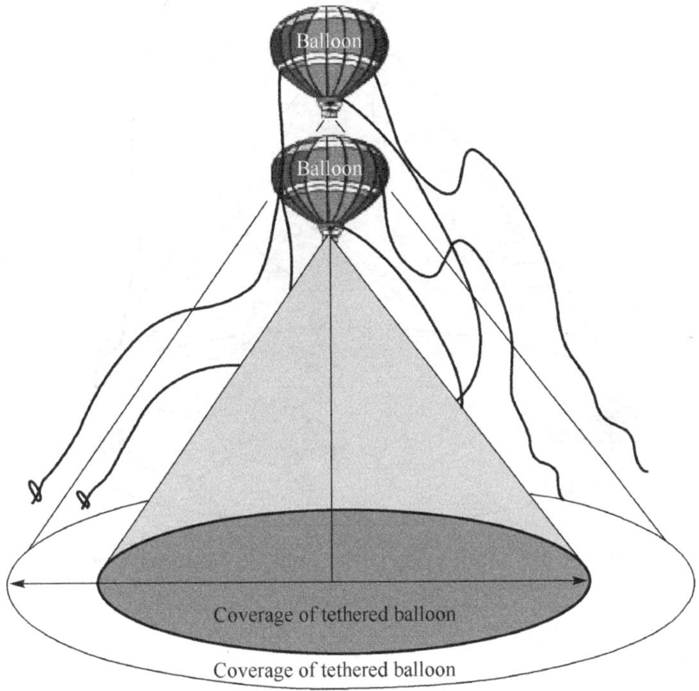

Figure 6-4 Changing the coverage of the tethered balloon according to the size of the disaster impact area

Disaster destroys all of the communication and electric supply infrastructure. Therefore, the emergency communication system must take into consideration that no support can come via ground transport. Therefore, power transmission technologies such as time-varying electromagnetic fields have to be used.

Wireless transmission power is the electric power that transmits from the power source to the destination machine without discrete man-made conductors. Wireless power transmission is useful in case of hazards. The machine transmitter is connected to the source of power, and then using an electromagnetic field for power transmission across an intervening space to one or more receivers, it is converted back to electric power and used. In 2011, Komerath discussed the role of emergency in the overall lighter-than-air architecture of power beaming. Figure 6-5 illustrates the architecture power beaming transport from the tethered balloon. A power supply at the ground is used to generate the power and then send it up to the balloon through a tethered device. Afterward, power beams from the balloon to rescue and relief teams' receiver devices over the disaster area.

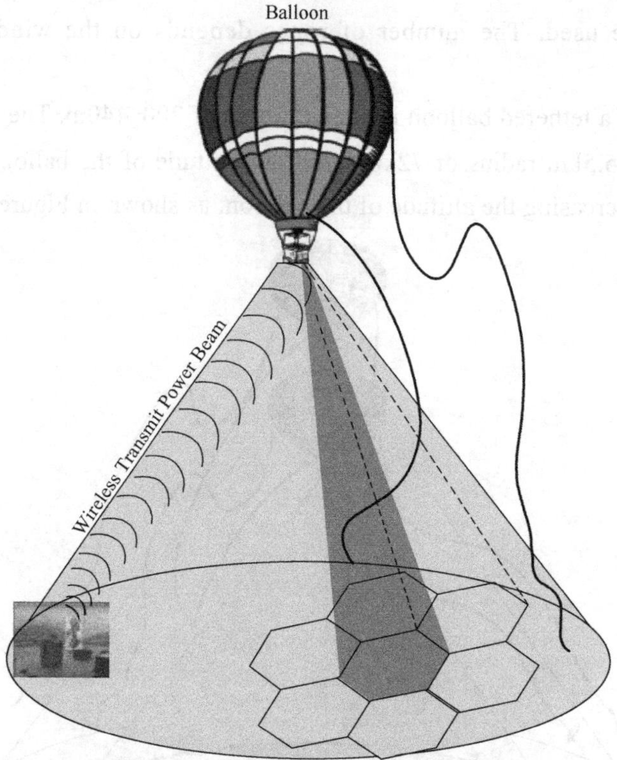

Figure 6-5　Delivering power from the tethered balloon

6.3　Airships

An airship, a lighter-than-air airship that overcomes its weight relies on the buoyancy generated by the light gas which is filled in the interior of the airbag, achieves rise, decline, hover or flight through power propulsion systems, boat-borne functional systems and flight control systems, etc. And it generally consists of airbags, empennages, pods, and propulsion units.

The airship can be divided into tropospheric airships and stratospheric airships according to their lift height. A tropospheric airship is an aerospace vehicle that primarily adapts to the tropospheric environment and performs multiple missions by carrying different mission loads. While the stratospheric airship is an aerostat that utilizes favorable conditions such as small wind speed near the height of the stratosphere to carry mission loads, relies on buoyancy to achieve liftoff, and keep stable long-term resident and controllable flight in a specific area.

The word airship is also used to denote all the air displacement vehicles that obtain buoyancy from the difference between the weight of the inflation gas within their hulls and the weight of the ambient atmosphere their bodies displace. This classification includes all airship-type vehicles with control and propulsion systems: traditional airships, unconventional airships, non-rigid, semi-rigid, and rigid airships, hybrid airships, heavy-lift air vehicles, high altitude airships, buoyancy assisted lift air vehicles, etc. Airships can be classified based on hull configuration (non-rigid, semi-rigid, and rigid), the way of producing vertical force (lighter-than-air, heavier-than-air, and hybrid), and payload capability (heavy-lift and medium-lift). The payload of traditional buoyancy air vehicles is usually less than 30 tons, while heavy-lift air vehicles can reach as high as 500 tons. Airships can also be divided into conventional types and unconventional types. In unconventional airships a major feature of the airship is distinctively different from the "conventional" type. The alteration could be attributed to shape and component design, lifting gas, unconventional lift method, payload, or power source. Generally speaking, conventional airships have a streamlined axisymmetric body, generate aerostatic lift by a hull with enclosed gas, have low payload capability, and use fuel as power source. All the other types are categorized as unconventional airships.

Airships have a great range of performance capability available to be exploited. One main advantage of airships is the low cost of energy consumption. Airships can hover for a long time without refueling and their operating costs are much lower than that of conventional fixed-wing airplanes or helicopters. Airships combine the advantages of both ships and airplanes. The speed of airships is higher than that of ships, their vibration levels are lower than that of airplanes, and they are not affected by sea state and a corrosive environment. Additionally, they can be boarded without the requirement for long runways. This enables them to transport heavy cargoes in remote areas. An airship transportation system causes low air and water pollution. It can meet challenging tasks for which airplanes and helicopters are not well-suited. Low noise and vibration levels as well as low vehicle accelerations provide an ideal platform for surveillance and patrol.

6.3.1 Main Attributes Offered by Airships

In this section, the attributes that can be demanded of an airship in performance of various roles can be listed while the assessments made below are generalized. Such refinement is considered to be outside the scope of this work, however. Listed here are the main attributes offered by airships in general.

1. Very High Endurance

As the most sought-after quality, some remarkable statistics appear. The World War I Coastal class, at 170000 cubic feet, would be considered small compared with the Skyship 600 of 1984, at 235000 cubic feet. Both types, on a normal patrol profile, could easily achieve 24 hours' endurance, whereas under favorable conditions both achieved 50 hours.

2. Speed

General statistics show a best cruise speed in then order of 25 to 35 n mile/h with maximum between 55 and 80 n mile/h, all speeds being size dependent.

3. Operating Altitude

Any airship must sacrifice gross volume to achieve altitude; thus a direct payload decrement occurs. In very rough-order terms, an airship required to take off at sea level and operate at

10000ft loses 25 percent of its lift gas volume; for 20000ft operation the volume loss rises to about 50 percent.

4. Crew and Equipment Environment

The low inertia forces and noise/vibration characteristics give a conventional airship a very good rating; some hybrids must be expected to be less acceptable.

5. Payload Volume

Structural weight considerations act in the airship's favor and space is very easy to provide. As a ratio with propulsion parameters, it becomes excellent compared with conventional aircraft.

6. Operating Costs

Direct operating costs are very low. If manual ground handling is used, together with mobile operations, the costs escalate very rapidly, however.

7. Safety

Very few viable statistics exist for airships in no-military operations. The history of operations with hydrogen lifting gas is dominated by the highly publicized loss of the Hindenburg, so it seems to be forgotten that the ship had made more than fifty commercial flights, including thirty-four ocean crossings. Put together with the operations of the two Graf Zeppelin airships, the statistics look much more favorable. Although the total passenger miles flown by these three airships is not available, a considerable figure was amassed. The Hindenburg alone, in just a year, carried 2800 passengers in a total of fifty-five flights and covered some 186000 miles. These figures give something in the order of nine million passenger miles. With the only fatalities being the thirty-five in the Hindenburg, the rate must have compared well with the other contemporary aviation statistics.

Civil commercial operations with helium as the lift gas are relatively few; the major accident record, however, is extremely good, there being no known fatalities. It seems the modern airship is inherently a very safe vehicle.

8. External Signatures

Because of the low mass, low propulsion energy requirement, and low structural density of airship, they can quite easily be designed to show very low external signatures. Radar, infrared, acoustic, magnetic, gravitational, and chemical emission signatures can all be reduced to low values.

9. Sensor Platform Capability

The airship has a proven historical capability as a platform for sensors. U.S. Navy airship in World War II carried radar to search for surfaced U-boats. The huge ZPG2 and ZPG3 airships carried a very large antennae, the 1.5 million cubic feet ZPG3W carrying a 40-foot span antenna inside the envelope when performing its distant early warning (DEW) function to provide a counter to the perceived trans-Arctic threat from Russian bombers. The very large sensors required for geographical survey are also quite easily mounted.

6.3.2 Conventional Airships

1. Non-rigid Airships

Figure 6-6 shows a typical configuration of classical non-rigid airships, called blimps as well. The shape of a non-rigid airship is sustained by a pressure differential between the lifting gas in the hull and the atmosphere. An envelope as the gas containment membrane encloses the lifting gas and the ballonets and provides protection from the environment. Ballonets are filled with air by blowers to maintain a fixed pressure inside as the temperature of the lifting gas or the airship altitude changes. Ballonets permit the envelope pressure to be controlled, and relative fullness of fore and aft ballonets is associated with pitch control. Adjustment of air volume in ballonets and gas volume in the envelope produces the change of buoyancy. The vertical portion of the car load is supported by an internal suspension system (adjustable catenary cable system), which is contained in the envelope and runs from the top of the envelope to the car. The principal function of the external suspension system attached to the bottom part of the envelope is to transfer the longitudinal components of the car loads into the envelope. The airship envelope fabric consists of laminated composite and is designed to withstand environment and flight loads. Lighter fabric can be used for ballonet materials because there is no flight load or environment exerted on the ballonet. The fabric used to make the envelope should have a high strength-to-weight ratio in order to reduce weight; low creep to maintain constant volume and shape; low permeability to ensure the purity of the lifting gas; high resistance to environment conditions to protect the airship from temperature, moisture, and ultraviolet radiation; and high fatigue and rupture strength to ensure the functionality of the envelope.

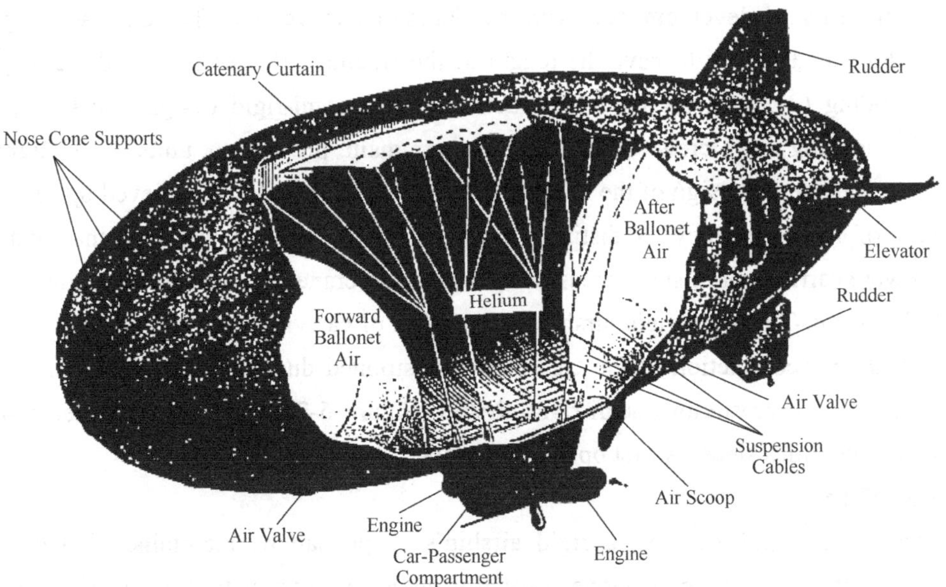

Figure 6-6 Typical non-rigid airship design

Non-rigid airships have simple structures and are easy to design, build, and maintain. In comparison with rigid airships, the fabrication cost of non-rigid airships is lower and the manufacturing time cycle is shorter. Non-rigid airships overcome the issue of weight penalty inherent in the use of rigid structures. Non-rigid configurations are especially suitable for small airships. There are drawbacks in building large non-rigid airships. A large amount of fabrics requires seaming of long length, vast working space, and special mechanical handling methods. Storage and shipping of helium for a large non-rigid airship may be a problem. Furthermore, the inflation of the envelope and the installation of empennage, nose structure, and gondola must be carefully dealt with due to the possible interaction with the pressurized hull.

2. Semi-rigid Airships

Semi-rigid airships have some characteristics of rigid airships and non-rigid airships. A rigid keel with an aerodynamic shape runs from nose to tail along the bottom surface of the air vehicle. In contrast to non-rigid configurations, the catenary suspension system plays a much reduced role and the keel supports the primary loads. This keel is used to eliminate the main function of the catenary curtain and evenly distributes the car weight along the airship's entire length. The interaction of keel and envelope may be partially favorable and partially unfavorable. The mutual support between keel and envelope is good for resisting and distributing the bending moments between them while the poor fit of keel to envelope causes them to act against each other and generate additional stress. Thus, an accurate characterization of the interaction of envelope and keel and their mutual effects is a crucial consideration for semi-rigid airship design. It can be anticipated that semi-rigid airships have weights between those of non-rigid airships and rigid airships, since the keel on the bottom acts like a structural load bearing member.

In recent years the development of semi-rigid airships has revived. The German Cargo Lifter model CL160 was designed to have the length of the Boeing 747s (852.8ft) and the height of a 27-story building (213.2ft). As the key structure of this semi-rigid design, the keel provided support for loading bay, crew cabin, load frame, main propulsion units, and flight deck. Innovative aerodynamic design of the heart-shaped profile of the CL160 achieved optimal lift and high levels of fuel efficiency. A distinctive feature of the CL160 was that the loading and unloading were carried out in small areas using a patented crane-like load frame while the airship remained in the air. This semi-rigid airship was capable of carrying heavy, large size goods of up to 50m length, but completion of the CL160 was postponed due to lack of funds in 2002. The Zeppelin NT-07 is a prestigious semi-rigid airship (Figure 6-7) which has provided more than 65000 passengers rides since its first operation in 1997.

3. Rigid Airships

In contrast to non-rigid airships, a rigid airship's shape can be maintained independent of envelope pressure because the envelope is usually supported by a metal framework, as shown in Figure 6-8. All external loads are carried by this lightweight structural outer shell. The external support structures are composed of a variety of transverse girders forming approximately circular

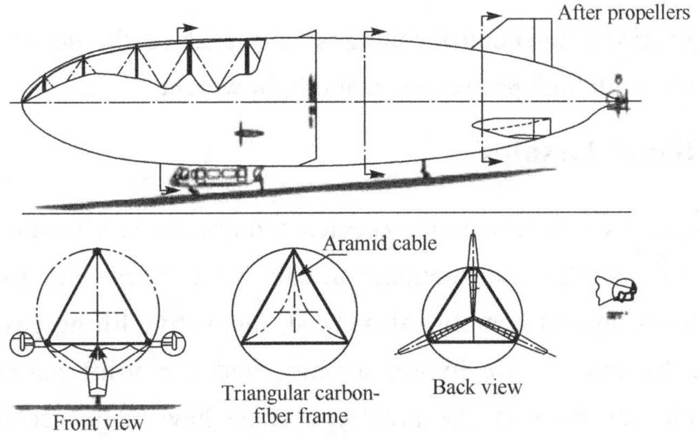

Figure 6-7 Zeppelin NT-07

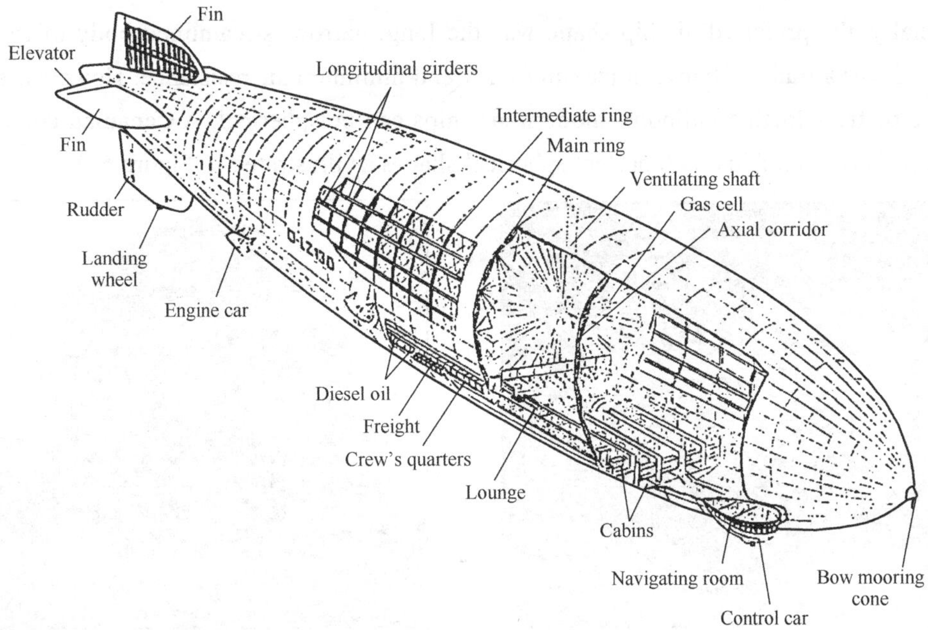

Figure 6-8 Typical rigid airship

frames and longitudinal girders running through the length. Transverse girders, usually made of aluminum, are connected by longitudinal girders, and are cross-braced with pre-tensioned metal wires for increased structural strength. Many gas cells containing lifting gas are placed between transverse frames. Gas compartmentalization of rigid airships increases safety and avoids sudden loss of substantial lift during an emergency. Lift adjustment due to altitude or temperature change can be accomplished by expansion and contraction of individual gas cells. The strength requirement of the envelope materials for rigid airships is lower than that of non-rigid airships since there are no large suspension system loads applied on the envelope. Rigid airships are usually constructed with a load- bearing frame, which allows them to accommodate all sizes and types of cargoes. Stress concentration of rigid airships is produced by the main elements of car, fins, and engines, which might be interconnected by internal structures. Recent advancements in new materials and superior connection techniques have enabled the design and construction of rigid airframe structures with high performance and light weight.

6.3.3 Unconventional Airships

In the last half-century, there has been an unexpected and dramatic renaissance in the development of novel buoyancy air vehicles. Considerable attention has been paid to the unconventional aspects of unique shapes, hybrid operational method, innovative lifting gas, and heavy payload capability. Of particular interest are hybrid airships and heavy-lift air vehicles. Preliminary developments of some programs at the prototype stage have been accomplished, and many ongoing projects are heading towards a promising and fruitful direction.

1. Spherical Airships

Traditionally, the preferred airship shape was the long, narrow, streamlined body of revolution, which achieves a tradeoff between maximum lift and minimum air resistance. However, adopting the shape of free-floating balloons, modern air ships can be spherical. A Canadian company has built six prototype airships with perfect spherical shapes as illustrated in Figure 6-9.

Figure 6-9 Spherical airships of a canadian company

This kind of airship is not equipped with control fins and an external gondola. Instead, the gondola is enclosed in the envelope at the bottom of the sphere and two engines are mounted at protruding wings outside the envelope. Though spherical shapes bring high aerodynamic drag compared with other configurations, they have distinctive merits. It is known that a spherical shape provides minimum surface area for a given volume among all the geometries. As the surface area is proportional to envelope weight, spherical shape generates maximum lift with minimum weight. Moreover, spherical shapes bring excellent features for operation and mooring: the airship does not need forward speed to land or take off as a conventional airship does; its spherical shape allows it to be moored by tying to the ground without using a mooring mast. Spherical shape and conical shapes can be combined to produce a distinct configuration. Figure 6-10 shows this geometrical design of non-rigid airships for rain forest exploration.

Experimental research demonstrated that the drag coefficient can be reduced by about 50% by placing a cone behind a sphere.

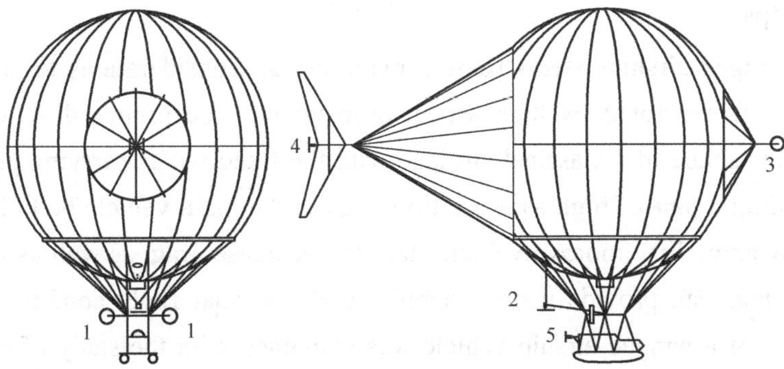

Figure 6-10 Schematic of a non-rigid airship for rain forest exploration

2. Lenticular Airships

Lenticular airships are easily affected by payloads during mooring in contrast with traditional airship bodies. The lenticular airship prototype called Alize was produced by the French LTA corporation in 2006 (Figure 6-11).

Figure 6-11 Lenticular airship

Lenticular airships have aerodynamic characteristics approaching those of wings and therefore make it possible to compensate for accidental overweighting (loss of helium, icing, etc.) through aerodynamic lift generation. The aerodynamic shape of lenticular airships is also helpful for flight maneuver control.

Double hull and multiple hull designs: Two or multiple conventional hulls of streamlined bodies can be joined together with or without connecting structures. This design achieves a reduction of overall length for a given volume of gas or an increase of gas compartment without an increase of overall length. Two large hulls can be connected by an inboard wing, which leads to increased aerodynamic lift and load capability. A novel type of double-hull design of airships is displayed in Figure 6-12. A double-hull configuration rather than a single hull reduces the lateral surface area and makes the airship less sensitive to lateral gusts. Double-hull designs were used for hybrid air vehicles.

3. Winged Airships

The concept of winged airships stems from airplane design considerations to take advantage of the aerodynamic lift generated by high aspect ratio wings. The proposed Ames Megalifter in Figure 6-13 has the shape of a classical airplane with the fixed wings carrying propeller turbines or turbofans. Adding a pair of high aspect ratio wings to the main vehicle body helps to produce substantial aerodynamic lift, improve vehicle stability, decrease drag, as well as increase payload capability. The wings can provide natural stability under normal flight conditions. In the recent decade, the design of a winged-airship vehicle was combined with the study of reliability, safety, and stability. Feasibility analyses, numerical simulations, and prototype fabrication were carried out.

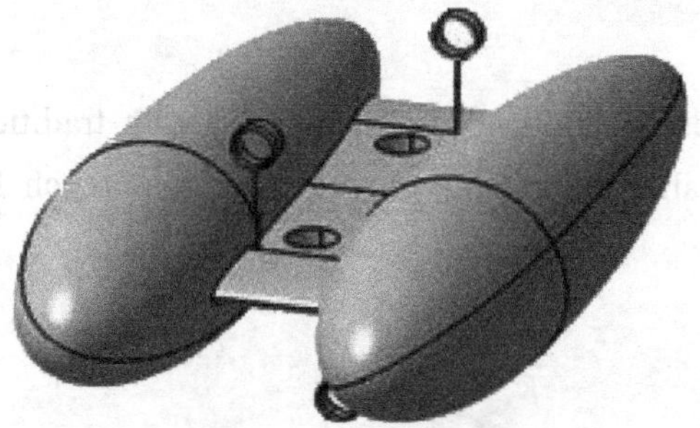

Figure 6-12 Double-hull design of unmanned airships

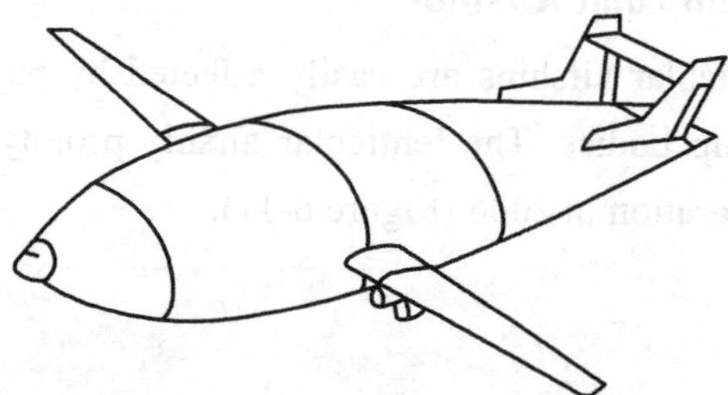

Figure 6-13 Winged-airship design

4. Hybrid Airships

Hybrid airships have been an active research area in airship history. The operation of traditional airships depends on the lighter-than-air condition. Hybrid airships combine the features of lighter-than-air and heavier-than-air vehicles and do not necessarily rely on conventional methods for lift generation. Hybrid airships derive the buoyancy partially from a lighter-than-air gas and partially from dynamic lift generated by shape and geometry. Hybrid air vehicles are usually found in combination with unconventional shape configurations. They may employ helicopter rotors, a wing-shaped lifting hull, a unique lifting body, or multiple hulls. Hybrid airships overcome the disadvantages of airplanes for long take-off and landing runways or of helicopters for large rotors. Hybrid air vehicles can carry significantly more payloads than conventional airships of similar size and are much less sensitive to weather effects.

An inboard wing used to connect two hulls has an effectively infinite aspect ratio of span to chord, which precludes the loss of lift due to tip flow. The load-carrying capability of this kind of hybrid airships depends on the volume of gas for buoyant lift, and on the inboard-wing, flight speed, and altitude for dynamic lift. The combination of both lifts allows for heavy loads to be carried. Hybrid air vehicles cannot be fully described by either airplane-derived or airship-derived relations. Kuhn et al. of the Technical University of Munich designed a demonstrator hybrid airship for ground observation which has the unique shape shown in Figure 6-14. Aerostatic lift and aerodynamic lift are used for energy efficient horizontal flight, while aerostatic lift and motor thrust are used for energy-efficient hovering. The semi-rigid structure was built out of sandwich beam structures and longitudinal rods. The inner structure was attached to the hull membrane through longitudinal rods and beams.

Figure 6-14 Schematic of hybrid airship

6.3.4 Stratospheric Airships

A stratospheric airship is a lighter-than air aircraft which can fly in the stratosphere with long endurance, high payload-to-weight ratio and low energy consumption. Generally, the stratospheric aerostat mainly includes the near space airship and high altitude balloon which are divided into two modes: station keeping and cruise with the wind (Figure 6-15). Figure 6-15(a) Station keeping: the propulsion subsystem is to provide enough thrust to overcome the drag of platform, then to remain stationary. Figure 6-15(b) Cruise with the wind: the platforms are directed by rising or descending into a layer of wind blowing in the desired direction. (In the stratosphere, there are many layers of wind, and each layer of wind varies in direction and speed.)

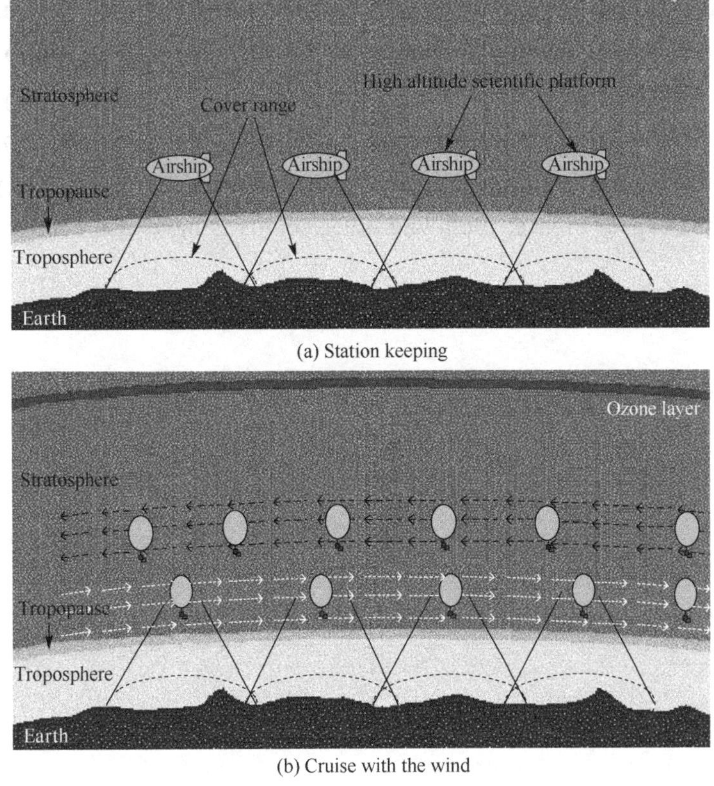

Figure 6-15 Comparison of two kinds of high altitude scientific platforms

It is ideally suited to provide potential applications especially for border patrol, homeland security, maritime and airborne surveillance, data and communications relay, and environmental research that require reliable and persistent station keeping capability. Meanwhile, as a new high altitude scientific platform (HASP), the stratospheric aerostat is required to provide scientific and technological investigations, including fundamental scientific discoveries that contribute to the understanding of the Earth, the solar system, and the universe.

The last 20 years have witnessed the projects establishment and researches of HASP in various countries, especially the developed countries in Europe and America, which greatly promote the development of HASP. In particular, with the rapid development of related subjects, the flight tests of these scientific platforms are more frequent recently.Researches on HASP are concerned on the energy and management, control system design, the performance test of airborne equipment and the capability of station keeping. Significant progress has been made during the flight tests. For example, NASA successfully released a super pressure balloon loading the COSI (the Compton Spectrometer and Imager) to explore the gamma-ray bursts, black hole, and the mysterious origin of the galactic positron and to study the birth and growth of the new galactic elements on 17 May 2016. Meanwhile, this high altitude balloon also carried an infrasound detector developed by Chapel Hill of the University of North Carolina to record the low frequency sound waves in the stratosphere. In addition, the airship team of Beijing University of Aeronautics and Astronautics has launched the China's first near space airship on 13 October 2015. The team achieved a number of key technologies such as high performance envelope material, flexible thin-film solar cells, high efficiency energy storage battery technology and integrated test 3. This near space platform was equipped with wideband communication, high resolution observation and spatial imaging and situational awareness systems to perform a variety of tasks. In addition, Japan, France, India and other countries have also conducted flight tests of HASP.

第6章 浮 空 器

6.1 介 绍

比空气轻的飞行器，通常被称为飞艇或飞船。随着新材料和新技术的发展，轻于空气的飞行器从热气球开始，逐步演变为气体填充的系留和非系留浮空器、飞艇和新型浮力飞行器。

浮空器是一种内部填充轻于空气的气体，且依靠或主要依靠空气静浮力来实现升空的航空器。浮空器主要分为气球和飞艇两大类。

在20世纪90年代中期，出现第一个采用了商用现成涂层材料的浮空器概念，满足了灵活性和保留氦气的要求。采用这种材料所设计的浮空器能进行充气和放气，飞行时能承载2～4人。浮空器的船体、鳍片、气囊等都采用了这种材料，鳍片能从船体部分填充氦气。

浮空器索具和硬件简单。承受的重量保持在飞机承载包络线内，因此承载重量的变化不会影响浮空器的俯仰角。一个连续运转的风扇使气囊保持在其失速压力下。一个用于氦气膨胀的快速拆卸接头安装在用螺栓固定在环组的板上，移除板可以排气。

系泊系统是一个安装在运输车后部的系绳绞车，配有发电机和滑环，为浮空器提供动力和信号。由于设计的紧凑性要求，在浮空器不飞行时，需要将其从运输车上卸下并转移到拖车上，以便在操作期间将运输车开走，同时还提供更大的存储能力。改进浮空器的鳍片形状和尺寸，以改善飞行特性，同时优化材料以改善其耐用性，并改进索具以提高其可靠性。

6.2 气 球

6.2.1 气球的分类

气球一般可分为系留气球和自由气球。其中，系留气球又可分为常规系留气球和载人系留气球两种，自由气球包括热气球和探空气球。

常规系留气球主要依靠充入气囊内部的比空气轻的气体产生浮力，克服其自身重量，并在空中漂浮，同时使用缆绳将其系留在地面阵地或牵引系留车上，利用地面绞车系统等设备控制其在大气中的漂浮高度来实现升降。目前，常规系留气球大多使用一种称为氦的安全惰性气体作为产生浮力的介质。常规系留气球系统主要由空中球体部分、系留缆绳组件、锚泊设施和地面综合保障设施等构成。根据其使用特点的不同，系留气球还可分为机动式、阵地式以及船载式系留气球(图6-1)。

载人系留气球与常规系留气球的工作原理基本类似。常规系留气球是为特定任务而设计的，而载人系留气球大多是娱乐性质的，因此对其飞行高度要求不高，但对飞行性能有更高的要求。载人系留气球一般由气球、地面设备和转向设备组成。载人系留气球下降时，主要通过地面绞车将气球收回至地面的降落平台，如图6-2所示。

图 6-1　我国第一艘实用型车载机动式系留气球

图 6-2　载人系留气球

6.2.2　系留气球技术

当地面通信基础设施在自然灾害中遭到破坏时，只有空间通信技术可以有效降低灾害的影响。因此，空间技术在灾后恢复过程中发挥着至关重要的作用。它们用于收集数据，以保护人类安全和减少经济损失。在灾难发生后，卫星通信是可靠的通信解决方案之一。然而，卫星通信有通信延误和发射成本较高的缺点。因此，考虑到卫星和无线通信系统的优点，使用空中平台是最佳解决方案。空中平台包括高空平台(HAP)、中高空平台(MAP)和低空平台(LAP)。

系留气球技术属于 LAP 系列，运行高度为地面以上 200～440m。系留气球的显著优点是部署成本低，传播延迟小，能快速部署并且有固定的工作站，尤其适合在灾害情况下使用。因此，在紧急情况下，使用系留气球技术对于自然灾害和人为灾害至关重要。同时，这些优点使其更适合快速部署的任务，使用户在灾区中使用手机进行紧急通信更加方便。因此，它是抢险救灾工作中的最佳解决方案，能够帮助救援队有效且高效地执行任务。系留气球装载有全球移动通信系统(GSM)、码分多址(CDMA)、特设、长期演进技术(LTE)、通用移动电信系统(UMTS)等，如图 6-3 所示。

Saeed Hamood Alsamhi 等教授研究了用于救援队应急通信服务解决方案的系留气球技术。系留气球是利用缆绳系在地上的气球，系留气球绳索长度不超过最大气球高度。为了保持气球的抗风稳定性，故使用了缆绳。缆绳的数量取决于风速预测和目标高度。

系留气球的运行高度为 200～440m。对于飞行高度为 440m 的气球，其提供服务的覆盖范围的半径为 5.5km，面积为 72km^2。其范围可以通过增加气球的高度来扩展，如图 6-4 所示。

灾难破坏了所有的通信和供电基础设施，因此必须考虑使用应急通信系统，但地面运输在灾后已经失去运载能力，必须使用电力传输技术，如变电磁场。

无线传输电力是指不通过分散的人造导体而直接从电源传输到目标用电设备的电力，在发生危险时起到极大的作用。发射装备连接到电源，然后使用电磁场通过中间站点向一个或

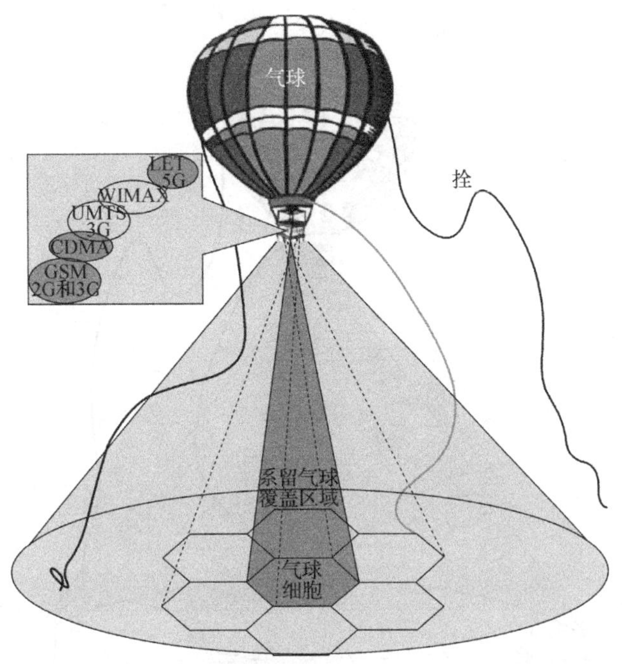

图 6-3 系留气球的有效载荷

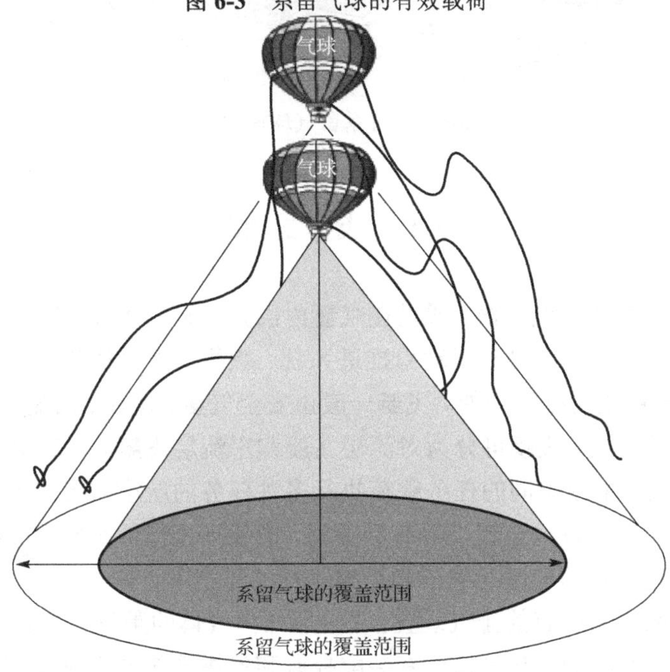

图 6-4 根据灾害影响区域的大小改变系留气球的覆盖范围

多个接收器进行电力传输，将其转换成可供使用的电力。2011 年，Komerath 提出了在紧急情况下功率发射技术在浮空器远程电力传输中的作用。图 6-5 展示了来自系留气球的架构功率发射传输。地面上的电源用于产生电力，再通过系留装置将其发送到气球。之后，系留气球的功率射线传送到灾区上空的救援队的接收设备。

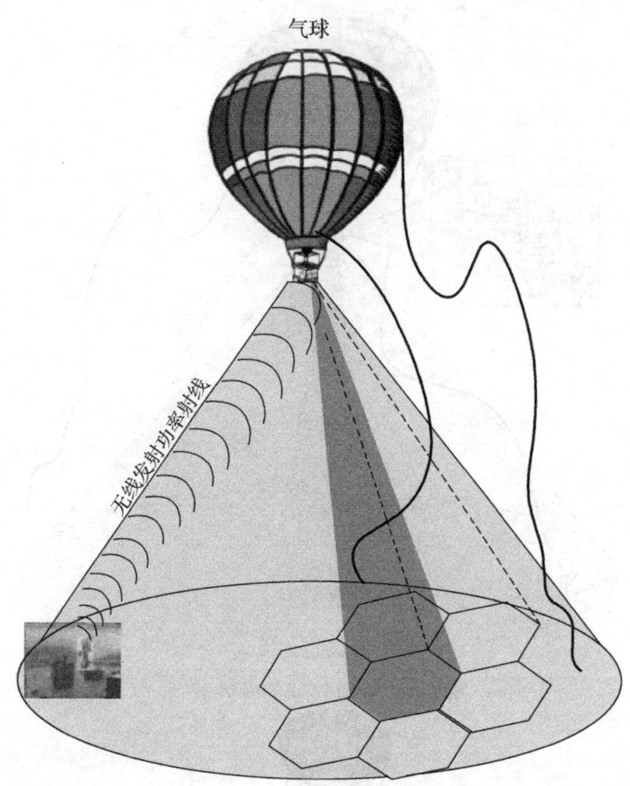

图 6-5　通过系留气球提供动力

6.3　飞　　艇

飞艇是一种比空气轻的航空器，通过在气囊内部充入轻质气体(通常为氦气)，依靠其产生的浮力来克服自身重量，同时利用动力推进系统、艇载功能系统和飞行控制系统等，实现升空、下降、空中悬停或飞行等功能。飞艇一般由安全气囊、尾翼、吊舱和推进装置等组成。

按其升空高度的不同，飞艇可分为对流层飞艇和平流层飞艇。对流层飞艇是指主要适用于对流层环境，并通过搭载不同的任务载荷执行多种任务的航空航天飞行器。而平流层飞艇是利用大气平流层高度附近风速较小等有利条件，搭载飞行任务所要求的载荷，依靠浮力升空，在特定区域实现稳定的长时间驻留和可控飞行的浮空飞行器。

飞艇一词也用来表示所有的排气装置，通过船体内气体的重量与其身体周围环境大气的重量之间的差异来获得浮力。它包括所有装配控制和推进系统的飞艇型装置：常规飞艇，非常规飞艇，非刚性、半刚性和刚性飞艇，混合飞艇，重型飞行器，高空飞艇，浮力辅助升力飞行器等。飞艇可以根据船体结构(非刚性、半刚性和刚性)产生升力(轻于空气、重于空气和混合动力)的方式以及承载能力的大小(重型升力和中等升力)来进行分类。传统浮力飞行器的负载能力通常小于 30t，而重型飞行器的负载能力可高达 500t。飞艇还可以分为常规飞艇和非常规飞艇。在非常规飞艇中有一个主要特征与“常规”飞艇明显不同，这种差异可以归因于形状和部件设计、提供升力的气体、非常规提升方法、有效负载或电源。一般而言，常规飞艇具有流线型的轴对称船体，通过具有封闭气体的船体产生静升力，负载能力较低，并且使用燃料作为动力源。除此之外，其他类型的飞艇都是非常规飞艇。

飞艇具有很多优点，可供开发利用。一个主要优点是能耗低，可以在没有加油的情况下长时间悬停，其运营成本远低于常规固定翼飞机或直升机。飞艇结合了船舶和飞机的优点。飞艇的速度比船舶快，其振动水平比飞机低，并且不受海况和腐蚀性环境的影响。同时，飞艇起飞时不需要借助长跑道，这使其能够往偏远地区运输重型货物。飞艇运输系统导致的空气和水污染较低，可以完成飞机和直升机不适合的具有挑战性的任务。低噪声和低振动水平以及低加速度为监控和巡逻提供了一个理想的平台。

6.3.1 飞艇的主要特性

本节列出了飞艇在实际应用中体现的主要特性。这些特性只是概括性的，读者可以根据特殊类型和实际价值来做更精细的评估，这里只给出普遍的特性指标。

1. 高耐久性

高耐久性是飞艇最出色的性能。例如，第一次世界大战中的“海岸”级飞艇的量级是170000ft^3，它和1984年235000ft^{3}的天舟600型飞艇属同一个量级，这两种飞艇在执行普通巡逻任务时都可以持续24h，在环境条件好的情况下可以达到50h。

2. 速度

根据一般统计显示，飞艇的最佳巡航速度为25～35海里/小时，最高航速为55～80海里/小时，所有这些指标都是由飞艇体积大小决定的。

3. 飞行高度

任何飞艇要想升到一定的高度都必须牺牲膨胀浮升气体的体积，从而直接减少载荷。可以这样粗略估算：要升高到10000ft高度，在海平面就要减少25%的浮升气体体积，要达到20000ft则需要减少50%。

4. 人机环境

常规飞艇有很好的人机环境，如惯性力弱、噪声低、振动小，而混合动力型飞艇的情况相对差一点。

5. 有效载荷

若仔细考虑结构重量方面的问题，则很容易留出足够的承载空间；若以推进力参数来看，飞艇较飞机而言有很大优势。

6. 运行成本

飞艇的直接运行成本很低。如果地面上使用人工操作，结合机动作业，成本就会大大增加。

7. 安全性

非军事用途的飞艇，其运行的数据非常少，历史过分放大了氢气时代的兴登堡号灾难，而忘了它曾经完成过50次以上的商业航行，包括34次横越大洋。如果连同另外两艘齐柏林伯爵飞艇的统计数字一起分析，结果会更加出色。虽然这三艘飞艇确切的总乘客里程数无法统计，但也是相当可观的，单是兴登堡号，仅一年就航行55次共186000英里，共运送乘客2800名，拥有约900万英里的乘客里程数，而在兴登堡事故中死亡的乘客仅35人，与同时代的其他航空统计数据相比，这已经是相当安全的了。

充氦飞艇的商业运营案例相当少，根据记录，最大的一次事故也没有造成人员伤亡，这样看来现代飞艇是一种非常安全的飞行器。

8. 外部识别信息

因为飞艇质量轻，不需要很大的推进力，且结构密度很小，所以它的外部识别信息可以设计得很微弱。雷达、红外线、声、磁、重力、化学排放信号都可以降到低值。

9. 传感器平台性能

历史已经证明飞艇可以作为空中传感器平台。第二次世界大战中美国海军利用飞艇搭载雷达来探测浮出水面的德国潜艇。巨大的 ZPG2 和 ZPG3 型飞艇装载了一个很大的天线，1500000ft^3 的 ZPG3W 在艇囊内装有一个跨度 40ft 的天线，用于远程预警(DEW)，可以发现横越北极的俄国轰炸机。在飞艇上还可以很方便地装载用于地理测量的超大传感器。

6.3.2 常规飞艇

1. 非刚性飞艇

图 6-6 显示了典型非刚性飞艇的设计，非刚性飞艇也叫软式飞艇。非刚性飞艇的形状由船体中的浮升气体与大气之间的压力差来维持。作为气体保护膜的飞艇外壳包裹着气体和气囊，使其免遭环境的破坏。鼓风机向气囊填充空气，以在提升气体的温度或飞艇高度变化时保持内部压力。飞行员通过气囊来控制压力，并且前后气囊的相对填充度与飞艇俯仰控制相关。气囊中的气量和气体体积的调节会对浮力的变化产生影响。飞艇负载的垂直部分由内部悬架系统(可调节悬链线缆系统)支撑，该系统包含在整个飞艇封套中并从封套顶部延伸到底部，附接到封套底部的外部悬挂系统的主要功能是将飞艇负载的纵向部件传递到封套中。飞艇封套织物由层压复合材料制成，用于承受环境和飞行载荷。较轻的织物可以用在不承受飞机载荷的气囊上。用于制造外壳的织物应具有高的强度重量比以

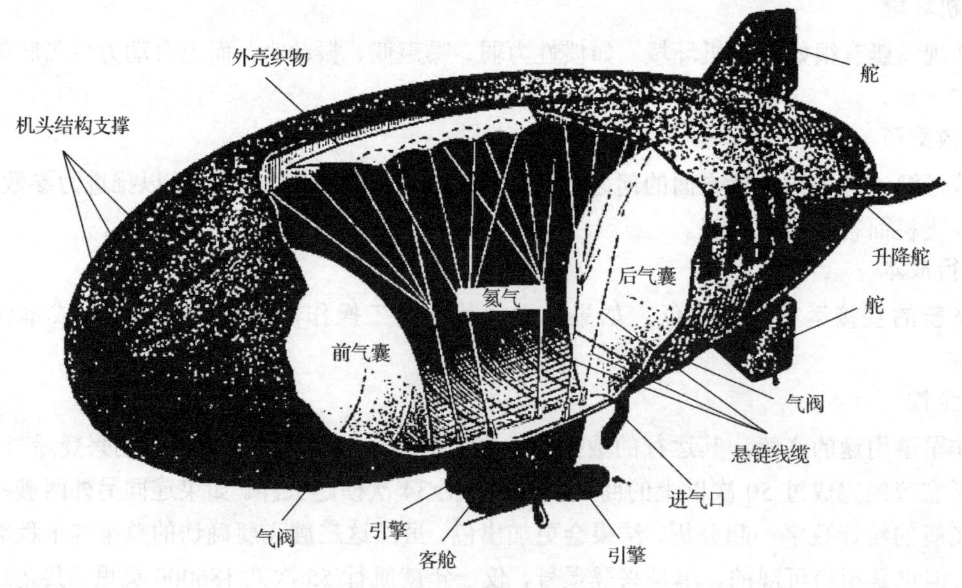

图 6-6　典型非刚性飞艇设计

减轻重量；低蠕变以保持恒定的体积和形状；低渗透性以确保提升气体的纯度；高环境条件耐受性以保护飞艇免受温度、湿度和紫外线辐射；高疲劳和断裂强度以确保外壳的寿命。

非刚性飞艇结构简单，易于设计、制造和维护。与刚性飞艇相比，非刚性飞艇的制造成本较低，制造时间周期较短。非刚性飞艇克服了使用刚性结构所固有的重量损失问题，其配置特别适用于小型飞艇。但在建造大型非刚性飞艇方面存在缺陷。在建造时有大量的织物需要长缝合，并且需要较大工作空间和特殊的机械处理方法。同时，用于大型非刚性飞艇的氦气的储存和运输可能是个问题。此外，由于可能与加压船体相互作用，必须小心处理气囊的膨胀以及尾翼、机头结构和吊舱的安装问题。

2．半刚性飞艇

半刚性飞艇具有刚性飞艇和非刚性飞艇的某些特征。具有空气动力学形状的刚性龙骨沿着飞行器的底部表面从头部到尾部延伸。与非刚性配置相比，悬链系统的作用大大减小，龙骨支撑主要载荷。这种龙骨承担主要载荷，并沿着飞艇的整个长度均匀分布重量。龙骨和外壳的相互作用可能既有益处也有不利因素。龙骨和外壳之间的相互支撑有利于抵抗和分配它们之间的弯矩，而龙骨与外壳的不良配合使它们相互作用并产生额外的应力。因此，准确计算包括龙骨和外壳的相互作用及其相互影响是半刚性飞艇设计的关键考虑因素。可以预期的是，因为底部的龙骨起到结构承载构件的作用，所以半刚性飞艇的质量介于非刚性飞艇和刚性飞艇之间。

近年来，半刚性飞艇的发展恢复了活力。德国型号为 CL160 的货物升降机的设计长度与波音 747s 的长度相当(852.8ft)，高度与 27 层楼（213.2ft）相当。龙骨作为这种半刚性设计的关键结构，为装载舱、乘员舱、负载框架、主推进装置和驾驶舱提供支撑。CL160 的空气动力学设计实现了最佳提升和高水平的燃油效率。CL160 的一个显著特点是使用专利的起重机式装载架在小区域内进行装载和卸载，而飞艇仍留在空中。这种半刚性飞艇能够承载长达 50m 的重型大型货物，由于 2002 年资金短缺，CL160 的竣工被推迟。齐柏林 NT-07 是一艘久负盛名的半刚性飞艇(图 6-7)自 1997 年第一次运营以来，已乘载了 65000 多名乘客。

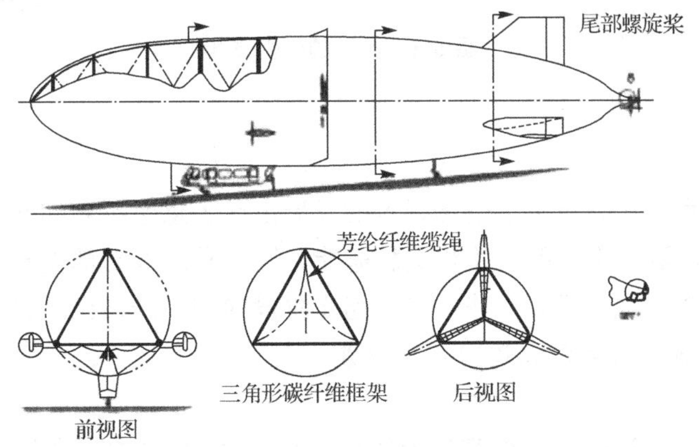

图 6-7 齐柏林 NT-07

3．刚性飞艇

与非刚性飞艇相比，因为刚性飞艇外壳通常由金属框架支撑，所以它可以不依靠外界支撑而保持其形状，如图 6-8 所示。所有外部负载均由这种轻质结构外壳承载。外部支撑结构由各种横梁组成，这些横梁形成接近圆形的框架和贯穿该长度的纵梁。横梁通常由铝制成，通过纵梁连接，并用预张紧力的金属丝交叉支撑，以增加结构强度。许多含有浮空气体的气室放置在横向框架之间。刚性飞艇的气体分隔增加了安全性，并避免了在紧急情况下突然失去大量升力。由于高度或温度变化引起的升力变化可以通过单个气室的膨胀和收缩来调节控制。刚性飞艇的蒙皮材料的强度要求低于非刚性飞艇的强度要求，因为在蒙皮上没有施加大的悬挂系统载荷。刚性飞艇通常采用承重框架设计，可以容纳各种尺寸和类型的货物。刚性飞艇的应力集中是由艇体、翼片和发动机产生的，这些元件通过内部结构相互连接。新型材料方面的进步和先进的连接技术使得刚性机身结构的设计和构造具有性能高和重量轻的特点。

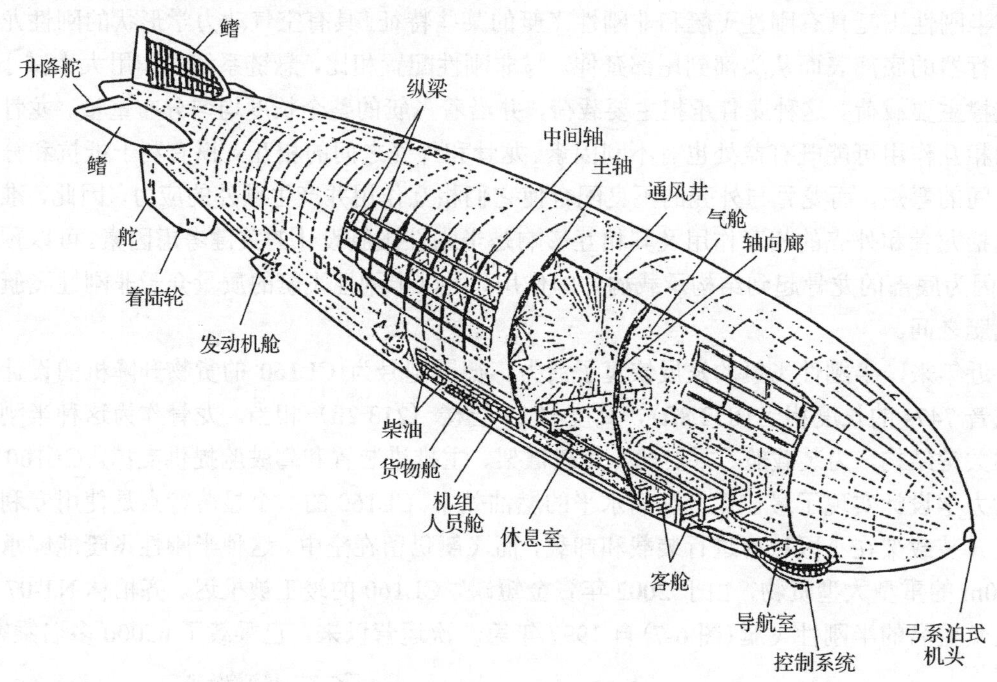

图 6-8　典型的刚性飞艇

6.3.3　非常规飞艇

在过去的半个世纪中，新型浮空飞行器的发展出现了意想不到的戏剧性的复兴。独特形状、混合操作方法、创新的起重气体和重载荷能力等非常规方面都受到了大量的关注。特别是混合动力飞艇和重型飞行器。一些项目在起步阶段的初步开发已经完成，许多正在进行的项目正朝一个具有前景和富有成效的方向前进。

1．球形飞艇

通常飞艇的形状优选的是长而窄的流线型旋转体，它在最大升力和最小空气阻力之间取得平衡。然而，采用自由浮动气球的形状，现代飞艇依然可以设计为球形。加拿大某公司建造了六艘具有完全球形的飞艇，如图 6-9 所示。

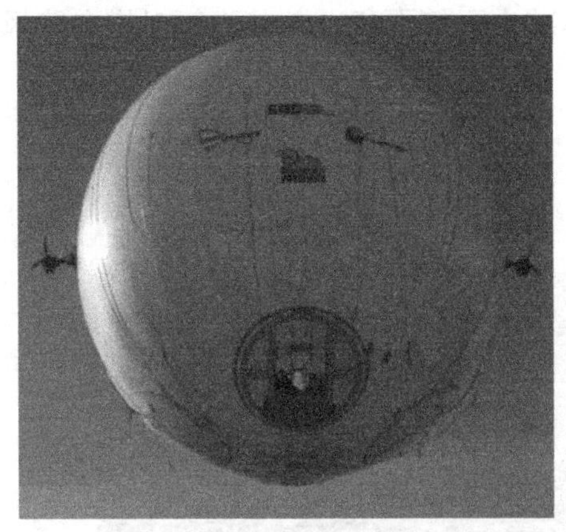

图 6-9 加拿大某公司的球形飞艇

这种飞艇没有配备控制鳍和外部吊舱。相反，吊舱被封闭在球体底部的外壳中，两个发动机安装在外壳的突出翼上。与其他配置相比，球形虽然带来了很高的空气阻力，但它们具有独特的优点：在给定体积的所有几何形状中球形的表面积最小。由于表面积与包络重量成比例，球形形状以最小重量产生最大升力。此外，球形形状为操作和系泊带来了优异的性能：飞艇不需要像常规飞艇一样对着陆和起飞的前进速度有要求；球形允许它在不使用系泊桅杆的情况下系在地上从而完成系泊。球形和圆锥形可以组合生成一个独特的配置。图 6-10 显示了用于雨林勘探的非刚性飞艇的几何设计。研究表明，在球体后面放置一个锥体，可以将阻力系数降低约 50%。

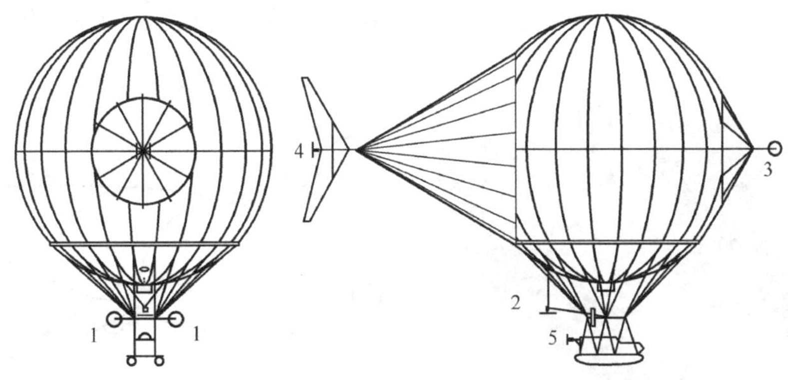

图 6-10 用于雨林勘探的非刚性飞艇示意图

2. 透镜状飞艇

与常规飞艇机身相比，在系泊过程中，透镜状飞艇很容易受有效载荷的影响。Alize 的透镜状飞艇是由法国 LTA 公司于 2006 年生产的，如图 6-11 所示。

透镜状飞艇具有接近机翼飞行器的空气动力学特性，因此它可以通过气动升力产生来补偿意外超重(氦气损失、结冰等)。透镜状飞艇的空气动力学形状也有助于飞行机动控制。

双层船体和多个船体设计：无论有没有连接结构，两个或多个流线型的传统船体都可以连接在一起。该设计在不增加船体总长度的情况下，可以减少给定的气体体积或者增加气室。两个大型船体可通过内侧机翼连接，从而提高空气动力升力和负载能力。一种新型的飞艇双壳设计如图 6-12 所示。双壳结构相比单个船体减少了侧面积，并使飞艇对横向阵风不那么敏感，该设计用于混合动力飞艇。

图 6-11 透镜状飞艇

3．飞翼飞艇

飞翼飞艇的概念源于飞机设计考虑，以利用高展弦比机翼产生气动升力。图 6-13 中提出的 Ames Megalifter 具有经典飞机的形状，其具有搭载螺旋桨涡轮机或涡轮风扇的固定机翼。在主结构上增加一对高展弦比的机翼有助于产生大量的空气动力升力，提高车辆稳定性，减少阻力，并增加承载能力。在正常飞行条件下，机翼可以提供自然稳定性。近十年来，飞翼飞艇的设计与可靠性、安全性和稳定性的研究相结合，人们进行了可行性分析、数值模拟和原型制造。

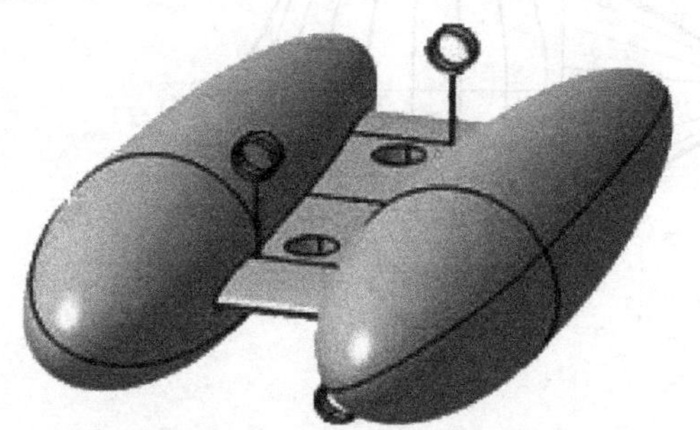

图 6-12 无人驾驶飞艇的双壳设计

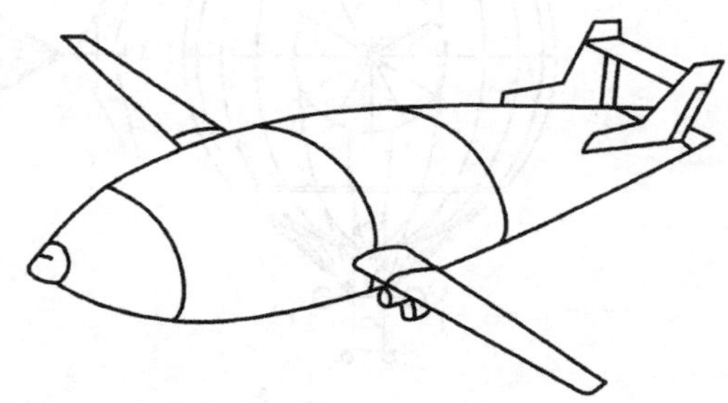

图 6-13 飞翼飞艇设计

4．混合动力飞艇

混合动力飞艇一直是飞艇历史上一个活跃的研究领域。常规飞艇依据密度比空气小的原理飞行。混合动力飞艇结合了轻于空气和重于空气装置的特征，并不一定依赖于传统的升力产生方法。混合动力飞艇一部分由轻于空气的气体产生浮力，另一部分由特定的几何形状(如机翼)产生升力。混合动力的设计通常与非常规形状配置相结合，可使用直升机旋翼、变翼型机体、特殊的升力产生结构或多船体的设计形式。混合动力飞艇克服了飞机长距离起飞和着陆必须使用跑道的缺点。与类似尺寸的常规飞艇相比，混合动力飞艇可以承载更多的有效载荷，并且对天气影响的敏感度要低得多。

用于连接两个船体的内侧机翼具有有效的无限跨度与弦长宽比，这种设计减小了由于机翼尖端湍流而导致的升力损失。这种混合动力飞艇的承载能力取决于浮力升力的气体体积，以及内侧机翼、飞行速度和动态升力的高度。双承载机体的设计允许承载重载。混合动力飞艇不能用飞机衍生的或飞艇衍生的关系来完全描述。慕尼黑技术大学的库恩等设计了一款用于地面观测的示范混合动力飞艇，其具有图 6-14 所示的独特形状。空气静力升力和气动升力用于节能水平飞行，而空气静力升力和电动机推力用于悬停。半刚性结构由夹层梁结构和纵向杆构成。内部结构通过纵向杆和梁连接到船体膜上。

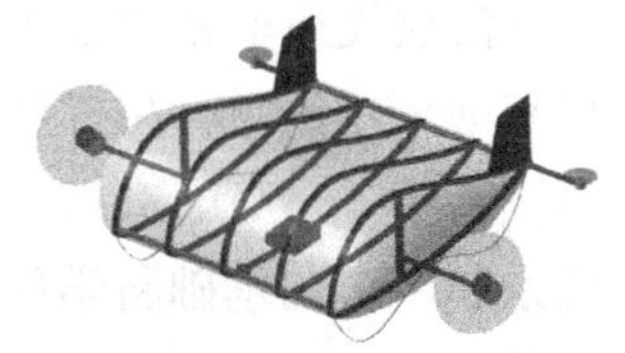

图 6-14 混合动力飞艇的示意图

6.3.4 平流层飞艇

平流层飞艇是一种比空气轻的飞行器，可以在平流层飞行，具有长续航能力、高承载重量比和低能耗的特性。通常，平流层浮空器主要包括近太空飞艇和高空气球，它们分为两种模式：站点保持和随风巡航(图 6-15)。站点保持：推进子系统提供足够的推力以克服平台的阻力，然后保持静止，如图 6-15(a)所示。随风巡航：平台通过上升或下降到一层向所需方向吹来的气流来引导，如图 6-15(b)所示。(平流层有很多层气流，每层气流的方向和速度都不相同。)

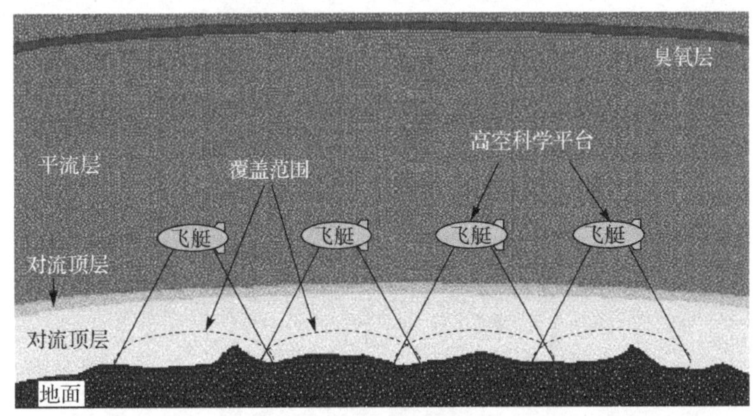

(a)站点保持

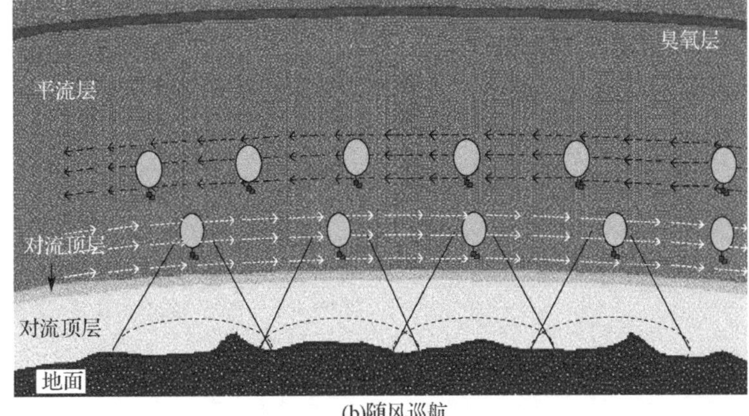

(b)随风巡航

图 6-15 两种高空科学平台的比较

平流层飞艇非常适合于提供潜在的应用，尤其是边境巡逻、国土安全、海上和机载监视、数据和通信中继，以及需要可靠和持久的站点保持能力的环境研究方面。同时，作为一个新的高空科学平台(HASP)，平流层浮空器可以提供科学和技术调查，包括有助于理解地球、太阳系和宇宙的基础科学问题。

近 20 年来，HASP 在各个国家，特别是欧美发达国家建立了研究项目，极大地促进了 HASP 的发展。随着相关学科的快速发展，高空科学平台的飞行试验最近更加频繁。关于 HASP 的研究涉及能量和管理、控制系统设计、机载设备性能测试和站点保持能力。它在飞行试验期间取得了重大进展。例如，2016 年 5 月 17 日，NASA 成功发布了装载 COSI(康普顿光谱仪和成像仪)的超高压气球，它可以探索伽马射线爆发、黑洞和银河正电子的神秘起源，并研究新星系的诞生和生长。同时，这个高空气球还带有一个次声探测器，由北卡罗来纳大学教堂山分校开发，用于记录平流层中的低频声波。此外，北京航空航天大学飞艇队于 2015 年 10 月 13 日推出了中国第一艘近太空飞艇。该团队完成了高性能蒙皮材料、柔性薄膜太阳能电池、高效储能电池等多项关键技术和综合测试。这个近太空平台配备了宽带通信、高分辨率观测、空间成像和态势感知系统，用来执行各种任务。此外，日本、法国、印度等国家也对 HASP 进行了飞行试验。

Chapter 7 Ground Effect Vehicles

7.1 Introduction

7.1.1 Definition of Ground Effect Vehicles

When the lower surface of the wing is close to the ground, airflow obstruction is formed, which increases the lift of the wing. This phenomenon is called the ground effect (GE) and is also known as the wing-in-ground (WIG) effect.

Ground effect aircrafts are vehicles that use the ground effect to provide support. They are sometimes called surface effect aircraft or wing-in-ground effect ships because of their typical flight conditions for the surface.

The WIG effect vehicle differs from a conventional airplane by the relatively small aspect ratio of the main wing, endplates (floats), special takeoff and alighting gear (takeoff or liftoff aids). The distinction from a conventional airplane can be seen from Figure 7-1, comparing the KM ekranoplan with the AN-225 (Mria) aircraft of similar size and weight. Ekranoplan was made by a distinguished Soviet engineer Rostislav E. Alexeyev who was the first in the world to develop the largest ground effect (GE) machine.

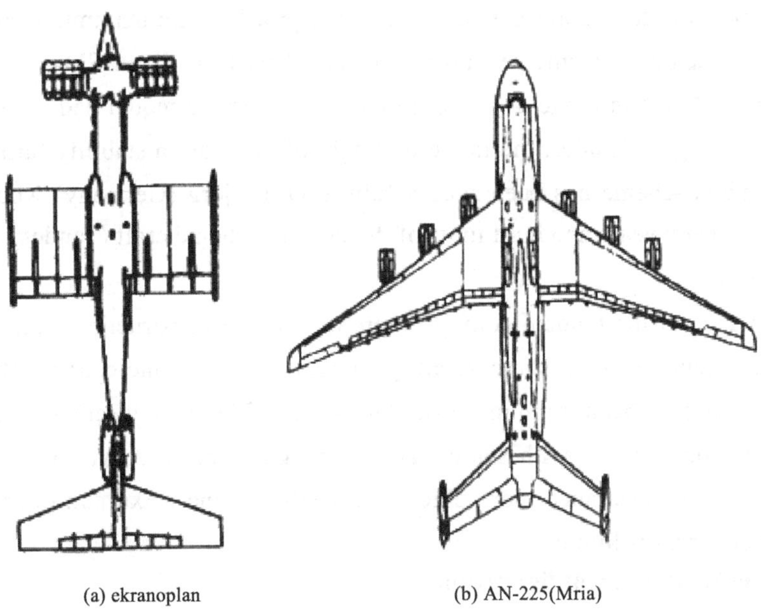

(a) ekranoplan (b) AN-225(Mria)

Figure 7-1 WIG versus airplane

The Soviet Military Encyclopedia adds to this list of distinctions of the ekranoplan the ‘raised location of the horizontal tail unit, beyond the limits of the influence of the ground and the wing wake, to ensure longitudinal stability’. Note that the latter feature may degenerate or completely vanish from some configurations such as ‘tandem’, ‘flying wing’ or ‘composite wing’. Contrary to the aircraft the WIG vehicles do not have to be hermetic. Conventional seaplanes versus WIGs: much larger aspect ratio and higher positioning of the main wing with respect to the hull, i.e. are less subject to the action of GE. Seaplanes (except Bartini’s VVA-14) are of airplane aerodynamic configuration. As compared to the hovercraft which is borne by a static air cushion, the WIG is supported by a dynamic air cushion that forms under the lifting wings at large speeds (RAM or chord-dominated GE) or/and by the wing-generated lift enhanced due to reduction of the down wash near the ground (span-dominated GE). While sharing some features with high-powered planning boats, the WIG is supported by dynamic pressure of the air whereas the planing boat is supported by the dynamic pressure of the water.

7.1.2 Classification of Ground Effect Vehicles

The wish to develop vehicles that exploit the GE which still have satisfactory longitudinal stability, has given birth to different aerodynamic configurations. In fact, the differences in configurations depend on the method of satisfying the longitudinal stability requirements. The basic configurations are as follows.

1. Tandem Configuration

The tandem configuration resolves the problem of stability by adjusting design pitch angles and the geometry of the fore and aft wing elements. This approach allows shifting the aerodynamic centers in a proper way for stability, while using wing profiles with maximum capacity to exploit the GE. The first tandem scheme self-propelled model was the 3-ton SM-1 launched in 1960. Although stable in a certain range of height-pitch parameters the model had a high takeoff speed and a ‘rigidity’ of flight. Besides, the range in height of the motion stability turned out to be too narrow. The tandem scheme has been successfully used by jörg (Germany) who developed this configuration for many years and built most of the tandem scheme craft (Tandem Aerofoil Boat—TAB) in the world.

The advantages of the tandem configuration are: simple construction, simple tuning of the configuration to secure a given static stability margin, effective one-channel (throttle) control, small span, i.e. length-to-beam ratio more similar to ships. The main disadvantage of this scheme is that it operates only in GE with static stability margin very sensitive to the combination of pitch angle and ground clearance. For vehicles of small size the maximum operational height is small and seaworthiness is limited.

2. Airplane-type Wing-tail Configuration

The airplane-type configuration features a large main wing moving close to the ground and a horizontal tail plane mounted on a vertical stabilizer outside the influence of the GE thus shifting the center of pitch downstream. The airplane scheme emerged from the Soviet R & D and construction work resulting in the creation of large ekranoplan of the first generation. Representatives of this scheme are 'KM', 'Orlyonok', 'Loon' and 'Strizh'.

The main advantages of this configuration are: large range of heights and height - pitch combinations for which the vehicle sustains stable flight (hence a capability to perform an emergency 'dynamic jump'), possibility to 'hop' and provide banking necessary for efficient turning maneuvers, possibility to efficiently apply power augmentation at takeoff. Large wing loadings leading to high-speed (this counterbalances the loss of the transport productivity due to low payload fraction). The disadvantages are: very large weight penalty for having a high-mounted sufficiently large tail unit (up to 50% of the area of the main wing), which contributes only insignificantly to the lifting capacity of the craft while adding additional viscous drag, relatively low lift-to-drag ratios (economic efficiencies) due to the large non-lifting area fraction as compared to high lift-to-drag ratio of the isolated main wing; large structural weight and, consequently, large empty weight fraction.

3. Flying Wing Configuration

The 'flying wing' configuration is characterized by remarkably reduced non-lifting components, and a very small (or absent) horizontal tail. Here the tendency is seen to convert the whole craft into a lifting surface, resolving the problem of longitudinal stability by special profiling of the lower side of the wing or/and by making use of an automatic stabilization/damping system.

The 'tailers' configuration of this type was proposed by Alexeev in the 1970s However, it was difficult to implement his ideas at that time and the scheme then was abandoned. Examples of 'flying wing'-type vehicles are: 'Amphistar-Aquaglide' (Soviet) and, a WISE vehicle under testing in Japan. Both of these crafts have natural stability due to smart profiling of the wing section. Formally, some other vehicles can be assigned to this type (e.g. KAG-3, Japan) although they do not have the 'flying wing'-type stability characteristics.

Advantages of the scheme are: efficient utilization of the vehicle to take maximum advantage of GE; low empty weight fractions, especially for vehicles of small aspect ratio. Disadvantages of this configuration are: supposedly low range of height-pitch combinations to achieve longitudinal stability (without the use of automatic control systems), relatively low operational flight heights, additional difficulties in providing structural integrity of a water-based all-wing vehicle; inefficient use of flaps which (additionally) may deteriorate the static stability of motion when they are employed improperly.

4. Composite Wing Configuration

The 'composite wing' configuration seeks to combine the advantages of the airplane configuration and the 'flying wing' configuration, thus achieving high takeoff efficiency when using power augmentation. A 'normal' composite wing has a central wing of small aspect ratio (centroplan) with endplates and side wings of high aspect ratio. It employs the idea of profiling the lower side of the main wing to reduce the tail unit. The overall aspect ratio of the 'composite wing' exceeds that of the main wing of the vehicles of the first generation and is by far larger than that of the tandem configuration as well as that of the existing 'flying wings' (less than 1.5). The latter property results in much higher lift-to-drag ratios and, in combination with S-shaping of the wing sections provides higher efficiency and range. The small aspect ratio of the centroplan provides maximization of the efficiency of the power-augmented takeoff. An example of the vehicle based on the 'composite wing' scheme is the MPE (Marine Passenger Ekranoplan) scaled series, e.g. the 450-passenger 400 ton ekranoplan MPE has a reduced tail area of about 27% of the main wing, and increased range of 3000km.

7.1.3 A Brief History of Ground Effect Vehicles

The earliest practical albeit unintentional utilization of GE belongs to the Wright brothers. The aviators encountered GE phenomena under the disguise of what was called a 'cushioning effect' or a 'pancake' landing. The transatlantic service of the seaplane Dornier DO-X demonstrated augmentation of the payload and range (1930–1931). Improved ride and handling qualities of conventional military aircraft (F105D, B-58, Avro Vulkan) even at distances exceeding five span lengths above the ground were regularly experienced.

The first purposefully designed GE vehicle was due to Kaario (Finland, 1935). His 'Aerosledge No. 8' featured a small-aspect ratio wing, leaning upon the skis (skegs) and a swiveling wing, directing the air propeller jet under the main wing. To provide additional static stability margin Kaario added two longitudinal rear beams with small stabilizing surfaces (Figure 7-2).

Figure 7-2 Kaario's Aerosledge No. 8

A precursor of the power augmentation system can be found in the Warner 'compressor' airplane (USA, 1928) (Figure 7-3). The design was based on a canard configuration and included two powerful fans forcing the air under a dome-like bottom of the vehicle. The Warner was the first to use separate takeoff and cruise engines.

The ram-wing concept was implemented by Troeng (Sweden, 1930s) (Figure 7-4). Particular features of Troeng's rectangular-wing vehicles were: ① enhanced static stability during takeoff with the help of special floats, ② use of a screw propeller, ③ use of a small hydrofoil at the trailing edge of the ram wing to ensure longitudinal stability in the design cruising mode.

Figure 7-3 Warner's 'Compressor' airplane

Figure 7-4 Troeng' s ram wing

Lockheed had been involved in WIG craft development since 1960. In 1963 a small two-seat boat with a wing fitted with endplates was launched (Koryagin). It had two bow hydroskis for better longitudinal stability. Lockheed is known to have studied a large WIG effect flying catamaran. The vehicle was to be stabilized and controlled by flap ailerons and a tail unit, comprising of vertical and horizontal rudders. The cargo was to be transported in the hulls and the wing. Later, Lockheed-Georgia studied a 1362 million lb wingship, which was designed as a logistics transport capable of transporting about 200 tons over 4000 nautical miles over an open ocean in a sea state 3 environment at a cruise speed of 0.40 Mach.

7.2 Projects of Ground Effect Vehicles

7.2.1 Projects Developed in China

Development and design of WIG effect craft in China was started in the China Ship Scientific Research Center (CSSRC) in 1967. Since then, during more than 30 years a total of nine small manned test vehicles have been designed and tested on lakes and in coastal waters. The XTW series was based on a wing-tail configuration with the main wing having forward sweep as in Lippisch designs (Figure 7-5).

Figure 7-5 XTW-1 vehicle

In 1996 the CSSRC reported developing the XTW-2, XTW-3 and XTW-4 WIG effect craft. A typical craft of this series is XTW-4 which was slightly modified from XTW-2 to comply with specific requirements from sea trials. This 20-passenger WIG effect ship was first tested on the Changjiang River in the autumn of 1999. The vehicle comprises: a major hull (float), the main wing supported by two minor floats, two vertical stabilizers carrying a high-mounted tail plane. To a certain extent the vehicle can be ascribed to wing-tail configurations. The main wing features the forward sweep. Two P & WC PT6A-15AG turboprop engines with MT's 5-bladed adjustable pitch propellers are mounted at the leading edge of the main wing. Thus, the slipstream is efficiently used to assist takeoff.

In early eighties another Chinese organization, MARIC, started developing what they called AWIG (Amphibious WIG). About 80 models were tested to study optimal wing profiles, configuration of the air channel, position of the bow thrusters, arrangement of the tail wing, etc. A self-propelled radio-controlled model of 30kg was tested on the lake in a suburb of Shanghai. As the model showed acceptable performance, MARIC proceeded to the development of the larger craft AWIG-750 with a maximum TOW of 745kg, length 8.47m, span 4.8m, height 2.43m (Figure 7-6). The power plant included internal combustion engines: two for lift and two for propulsion of the craft. Each engine drove a ducted thruster type DT-30 of 30hp rated power at 6000 r/min. The vehicle was able to take off in waves of 0.5m and had a maximum speed of 130km/h. It demonstrated the expected (amphibious) capability of passing from the water to the shore and back.

In 1995, the China State Shipbuilding Corporation commissioned the R & D for a 20-seat AWIG-751 under the name of 'Swan-I'. The vehicle which was completed by June 1997 had a TOW of 8.1 tons length-width-height dimensions of 19m×13.4m×5.2m and a maximum cruising speed of 130km/h in calm water. It had three aviation-type piston engines: two HS6E engines of 257kW, each for PAR lift and one HS6A engine of 210kW for propulsion. The PAR engines drove two bow ducted 4-bladed air propellers and the cruise engines drove a two-blade variable pitch propeller. As compared to the previous AWIG-750 it had several new features, including: increased span of the main wing, composite wing, combined use of guide vanes and flaps to enhance longitudinal stability, CHIBA composites to reduce structural weight.

Figure 7-6 AWIG-751

The tests confirmed overall compliance with the design requirements, but showed some disadvantages, namely, too long shaft drives of the bow propellers, lower payload and lower ground clearance than expected. The follow-on vehicle AWIG-751G (Swan-II) had increased dimensions, a modified PAR engines layout and an improved composite wing.

7.2.2 Projects Developed in Germany

Hanno Fischer, the former technical director of Rein-Flugzeugbau, set up his own company Fischer Flugmechanik and extended the Lippisch design concept to develop and build a 2-seat sports vehicle designated as Airfish FF1/FF2. Unlike X112 and the following X114, the Airfish was designed to fly only in GE. It was manufactured of GRP and reached a speed of 100km/h at just half the engine's power during tests in 1988. In 1990 Fischer Flugmechanik tested a 4-seat vehicle Airfish-3, which was 2.5 times heavier than Airfish FF2, flew at a speed of 120km/h and was able to cover a range of 370km (Figure 7-7). With a length of 9.45m and a width of 7.93m, the vehicle had an operational clearance ranging from 0.1 to 1m.

Figure 7-7 Airfish-3

A design based on the Airfish series formerly developed by Fischer Flugmechanik has re-emerged in Flightship 8 (FS-8 initially designated as Airfish 8) (Figure 7-8). The FS-8 was developed in Germany by Airfoil Development GmbH and made its maiden flight in the

Netherlands in February 2000. With its TOW of 2325kg, length of 17.22m, width of 15.50m and height of 4m. The Flightship-8 carries 8 people, including two crew. The wave height at takeoff is restricted to 0.5m, but when cruising the vehicle can negotiate 2m waves. FS-8 is made of FRP. With an installed power of 330kW it has a cruising speed of about 160km/h and a range of 365km.

Figure 7-8　Flightship 8

A larger Flightship-40 (FS-40) dubbed Dragon-Clipper is being designed for up to 40 passengers in the commuter version for an equivalent payload of 5 tons in alternative configurations. This larger craft has a length of 30m, and the wingspan of 25m can be reduced to 20m for onshore handling by folding winglets. The main construction material is aluminum, and the Pratt and Whitney turboprop-diesel engines developing 1000kW will increase the cruising speed to about 225km/h. Maximum takeoff wave height is 1.2m and increased wing span allows over-water operation in 4m seas. The originators of the FS-8 design Fischer Flugmechanik and Airfoil Development GmbH have recently announced a proposal to produce a new craft HW–20 combining WIG effect and static air-cushion technology. The design of HW–20 (Hoverwing) employs a simple system of retractable flexible skirts to retain an air cushion between the catamaran sponsons of the main hull configuration. This static air cushion is used only during takeoff, thus enabling the vehicle to accelerate with minimal power before making a seamless transition to true GE mode (Figure 7-9).

Figure 7-9　Hoverwing-20 with a static air-cushion liftoff system

7.2.3　Projects Developed in Russia

A composite wing configuration implies functional subdivision of the craft's lifting area into two parts: the one (central) taking advantage of the power augmentation mode, and the one (side wings) adding efficiency and longitudinal stability in cruise. Provision of stability in this case has three major ingredients: special profiling of the central part of the main wing, horizontal tail (albeit relatively small), appropriate geometry and position of the side wings. The designs, exploiting these features, are those of the MPE series (Designer General D. Synitsin), ranging in TOW from 100 through 400tons (Figure 7-10).

Figure 7-10 MPE-400

Ekranoplan Amphistar was developed and built by the company 'Technology and Transport' (Director and principal designer D. Synitsin) in 1995. In 1997 this vehicle was awarded the certificate of the Register of Shipping of the Russian Federation as a cutter on dynamic air cushion. The maximum TOW is 2720kg, its length-width-height dimensions are 10.44m×5.9m×3.35m. At cruising speed of 150km/h it has a range of up to 450km. Seaworthiness is about 0.5m. The turn radius at cruising speed is about 65 hull lengths. In water the turn radius is about a hull's length. A modified version of the vehicle has recently appeared under the name Aquaglide (Figure 7-11). Synitsin developed a scaled up series of Amphistar-Aquaglide-type vehicles (Figure 7-12).

Figure 7-11 Aquaglide-5 vehicle

Figure 7-12 Aquaglide-50 vehicle

7.2.4 Projects Developed in the USA

In the early 1990s, a US company named AEROCON developed a project Aerocon Dash 1.6 (Figure 7-13). This mammoth wingship had the following physical characteristics: TOW= 5000 tons, payload fraction of 0.3588, wing loading of 258lb/ft^2, cruise speed of 400 knots, cruise altitude of 12ft (3.66m).

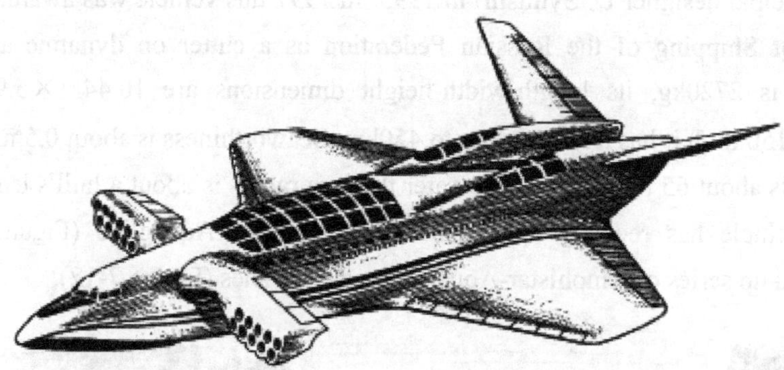

Figure 7-13 Aerocon Dash 1.6 vehicle

In recent years Lockheed Martin Aeronautical Systems(LMAS) investigated the development of what they call Sea-Based Aircraft. LMAS calls for a move to hybrid aircraft and search for appropriate hybrid solutions resulted in a family of designs. These include: seaplanes, floatplanes and WIG-like combined surface effects aircraft—SEA (Figure 7-14). SEA combines multiple surface effect technologies in a Sea-Based Mobility Hybrid Aircraft design—WIG, seaplane and hydroplane hull shaping. According to LMAS, such a concept is viable with the current aircraft technology, and would provide speeds up to 400 knots and a global range with 400 tons of payload.

As reported by Boeing Frontiers (online, September 2002, vol. 01, issue 05), a high-capacity cargo plane concept dubbed Pelican is being developed currently by Boeing Phantom Works (Figure 7-15). It has a large-aspect-ratio main wing, a wingspan of 500ft, a wing area of more than an acre, twice the dimensions of the world's current largest aircraft An-225, and it can transport up to 1400 tons of cargo.

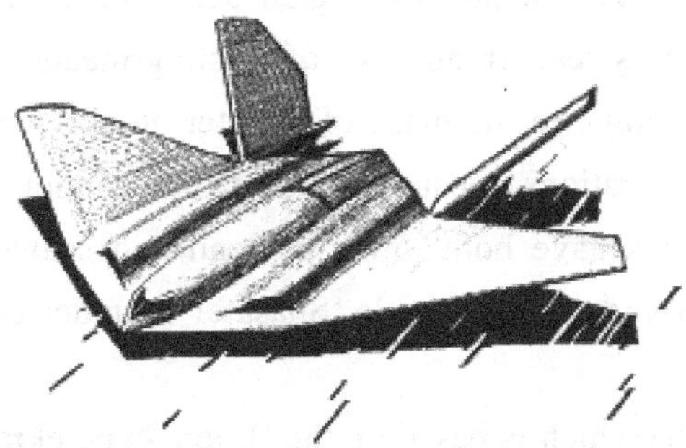

Figure 7-14 SEA (surface effect aircraft)

It has a long trans-oceanic range and can fly as low as 20ft above the sea (span-based relative ground clearance of the order of 20/500=0.04), but it is also able to fly at heights of 20000ft or higher, intended for commercial and military operators who desire speed, worldwide range and high throughput. As indicated by John Skoupa, senior manager for strategic development for Boeing advanced lift and tankers, 'The Pelican stands as the only identified means by which the US army can achieve its deployment transformation goals in deploying one division in 5 days or five divisions in 30 days anywhere in the world'. It can carry 17 M-1 main battle tanks on a single sortie.

Figure 7-15 Pelican

7.3 Application of Ground Effect Vehicles

7.3.1 Civil Application

According to a preliminary analysis, as reported by Belavin, Volkov et al. and Hooker, there exist encouraging prospects for developing commercial ekranoplan to carry passengers and/or cargo, to be used for tourism and leisure as well as for special purposes, such as search-and-rescue operations.

1. Search-and-rescue Operations

Memories are still fresh about the tragedies that happened with the nuclear submarine 'Komsomolets' on April 7, 1989 in the Norwegian Sea, and the nuclear submarine 'Kursk' on August 12,2000 in the Barents sea. An analysis of existing means of rescue on water shows that surface ships are unable to come to the place of disaster quickly enough, while airplanes cannot perform effective rescue operations because the airplanes cannot land close to a sinking ship. Even most modern seaplanes have both lower payload and seaworthiness as compared to the ekranoplan. The GE search-and-rescue vehicle 'Spasatel' is under construction at 'Volga' plant in Nizhniy Novgorod.

'Spasatel' (Figure 7-16) which is based on the 'Loon'-type ekranoplan, combines features of all known means of rescue on sea (search-and-rescue airplanes, helicopters, ships). Its cruising speed is expected to be in the range of 400–550km/h in GE, and up to 750km/h out of GE. Altitude when flying far from the underlying surface would be up to 7500m, and about 500m in searching mode. The vehicle can land and conduct rescue operations in waves up to 3.5m. It is capable of loitering in rough seas with wave heights reaching 4m. 'Spasatel' has a range of 3,000km, can operate autonomously for 5 days and is able to accommodate up to 500 people. Before a decision to develop 'Spasatel' had been taken several experiments on the available missile carrier 'Loon' have been performed to appraise the ekranoplan's capacity to serve as a rescue vehicle. These experiments showed that ekranoplan have some useful features justifying their use for rescue operations on the water. In particular, when drifting on water the vehicle is naturally brought to a position with its nose against the wind. As the vehicle's main wing is partially (with its aft part) immersed in the water, there forms a region of relatively calm water behind it.

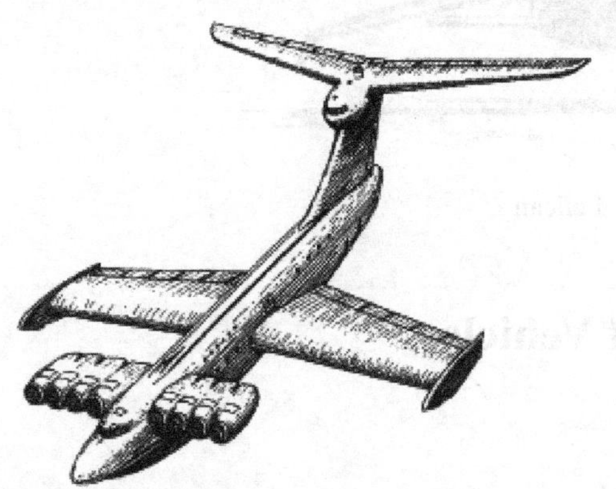

Figure 7-16 Search-and-rescue ekranoplan 'Spasatel'

2. Global Sea Rescue System

There is a worldwide concern to develop effective rescue measures on the high seas. Experience shows that it is very difficult if not impossible to provide timely aid at wreckages and ecological disasters at sea. Use of seaplanes is often limited because of unfavorable meteorological conditions, whereas use of helicopters is restricted to coastal areas. Until now, the main means of rescue (salvage) on water has been ships finding themselves accidentally near the disaster area and hardly suitable for this purpose.

A global sea rescue system is proposed, comprising 50 heavy weight ekranoplans, basing in 12 selected focal base-ports throughout the world. Each ekranoplan of the system is designed to have high take off/touch down seaworthiness, enabling its operation on the open sea during 95%

of the time year around. The cruise speed of each ekranoplan of the system is 400–500km/h and the radius of operation constitutes 3000–4000km. The vehicle can loiter for a long time upon the sea surface when seaborne at a speed of 15 knots. The rescue vehicle is supposed to bring to the place of disaster; a wide array of rescue means including rafts and self-propelled cutters and, possibly, helicopters and bathysphere.

3. Horizontal Launch of the Aerospace Plane

According to the project developed jointly by St. Petersburg State Academy of Aerospace Instrumentation and Musashi Institutes of Technology in Tokyo, Japan an unmanned self-propelled ekranoplan is supposed to carry, accelerate to almost half sound speed and launch a 600 ton rocket plane to a low earth orbit (horizontal launch) (Figure 7-17). Launching useful payloads into low earth orbit and expanding the functional capacity of the aerospace transport systems is one of the major tasks of the developers of new space projects for the 21st century.

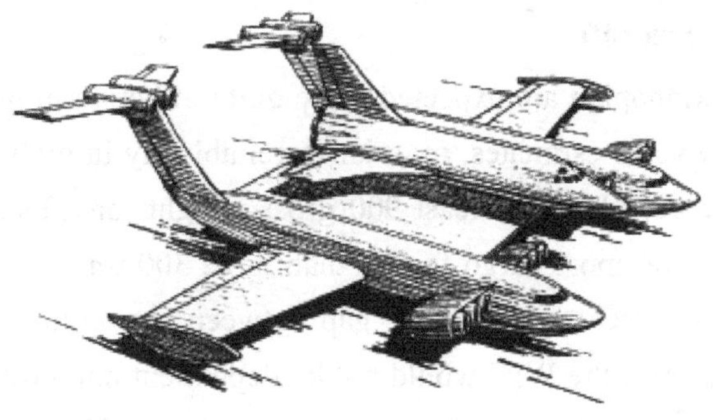

Figure 7-17 ekranoplan-rocket plane horizontal launch

7.3.2 Military Application

Analysis of known projects and future naval applications have confirmed that the above listed properties of ekranoplan together with their high surprise factor due to speed, low radar visibility, sea keeping capability, payload fraction comparable to similar size ships, dash speed feature and capacity to loiter afloat in the open ocean make them perfect multi-mission weapons platforms which can be deployed forward and operate from tenders.

Naval ekranoplan can be used as strike warfare weapons against land and seaborne targets, launch platforms for tactical and strategic cruise missiles, aircraft carriers and amphibious assault transport vehicles. Easy alighting at moderate sea states makes it possible to utilize ekranoplan as antisubmarine warfare planes capable of effectively deploying hydrophones or towed arrays. They can also be used in a wide variety of reconnaissance and transport roles. WIG effect vehicles could adapt themselves to an operational concept of anchorages all over the world to maintain a forward posture.

1. Amphibious Warfare

The speed, payload and low-altitude cruising capabilities of the WIG would enable devastating surprise assaults. It has also been noted by the analysts that the WIGs could have reached the Islas Malvinas from Britain in hours versus the days it took surface forces to arrive during the conflict. The major difficulty with PAR-WIG amphibious operations is the actual landing of men and equipment. Since reduced structural weight is a key factor enabling efficient WIG flight, the vehicle cannot be reinforced to allow beaching without deterioration of its cruise performance.

An example of an amphibious assault craft is the Russian ekranoplan 'Orlyonok'. Whereas the 'Caspian Sea Monster', notwithstanding such a threatening nickname, was not a combat vehicle, but just a huge flying test bed, 'Orlyonok' was the first ekranoplan, specially designed for military purposes. The vehicle with a combat load of 20 tons has a cargo compartment length 24m, width 3.5m and height 3.2m. To enable the embarkation-disembarkation of cargos of large dimensions and heavy military vehicles (e.g. tanks and armored carriers), 'Orlyonok' has a unique swing-away bow design.

2. Sea Lift

Ekranoplan are expected to be quite effective in providing a sealift function. However, as shown by some estimates, in order to reliably fly in high sea states, a trans-oceanic WIG would need to be very large, at least 900 gross weight tons. Even so it is estimated that one such WIG could deliver more cargo farther than three 300-ton C-5 aircraft, and do this while using 60% less fuel. The WIG would fill the gap between conventional air-lifters and slow surface shipping. Unlike aircraft, the WIG would not be dependent upon overseas bases. Yet, unlike ships, WIG sea-lifters would be fast, require no escorts, and would be invulnerable to torpedoes and mines.

3. Reconnaissance and Patrol

Maybe, the weakest mission application for large WIGs would be in reconnaissance or patrol. The limiting horizon resulting from low-altitude operation would greatly reduce radar or signal intercept range, and therefore area coverage, to the point where it might not represent a cost-effective use of the platform. Even in the strike warfare posture against ships, WIGs would require targeting information from other platforms.

第 7 章　地效飞行器

7.1　介　　绍

7.1.1　地效飞行器的定义

当机翼下表面靠近地面时，会形成气流阻塞，增加机翼的升力。这种现象称为地面效应(GE)，也称为翼地面(WIG)效应。

地效飞行器是利用地面效应提供支撑的运载工具。由于其典型飞行状况为贴水面飞行，因此也称为水面效应飞机或翼地效应飞机等。

地效飞行器与传统飞机的不同之处在于主翼、终板(浮子)、特殊起飞(起飞或升空辅助装置)和下降装置的纵横比较小。将 KM ekranoplan 与具有相似尺寸、重量的 AN-225(Mria)飞机进行比较，从图 7-1 中可以看出它与传统飞机的区别。(注：KM ekranoplan 也被称为“kaspian Monster”，中文译名为“里海怪物”，后面书中统称为“ekranoplan”。)ekranoplan 由苏联的杰出工程师 Rostislav E. Alexeyev 研发，它是世界上第一个最大的地效飞行器。

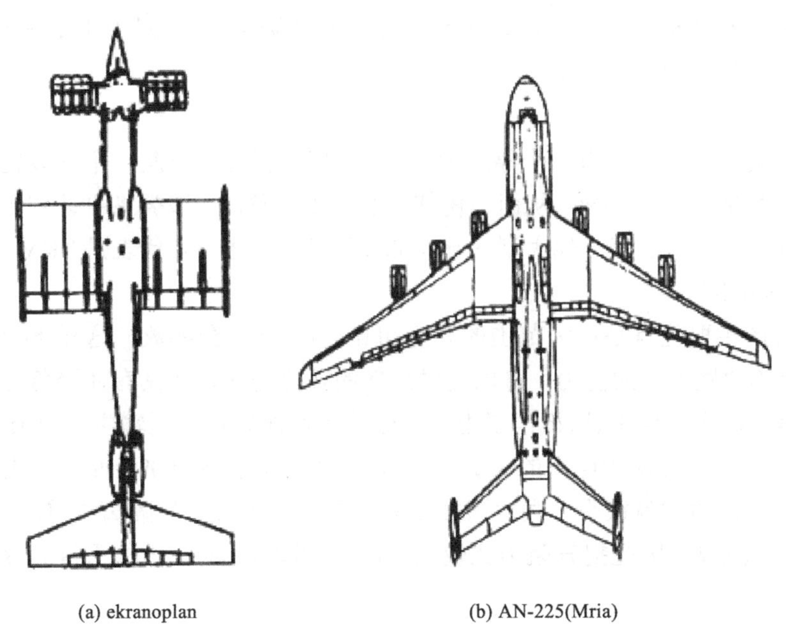

(a) ekranoplan　　(b) AN-225(Mria)

图 7-1　地效飞行器和普通飞机的比较

苏联军事百科全书增加了 ekranoplan 的区别，即“提高水平尾翼的高度，确保尾翼不受到地面效应与机翼尾流的影响，以达到提高操纵稳定性的目的”。机翼尾流的特征可能会从某些配置中退化或完全消失，例如“串联”、“飞翼”或“复合翼”。与飞机相反，地效飞行器不必是密封的。与地效飞行器相比，传统的水上飞机具有更大的纵横比，主翼相对于船体的定位更高，即更少受地面效应的作用。水上飞机(Bartini 的 VVA-14 除外)具有飞机空气动力学配置。与由静态气垫承载的气垫船相比，地效飞行器由动力气垫支撑，动力气垫在升力机翼下以大速度(舷梯或弦主导地面效应)形成或(和)由机翼产生的升力增强，这是由于减少了靠近地面的下冲(跨度主导的地面效应)。虽然与大功率规划船有一些相同特征，但地效飞行器受到空气动态压力的支持，而滑行船则由动态水压支撑。

7.1.2 地效飞行器的分类

开发和利用地面效应的飞行器并使其具有较好的纵向稳定性，这就产生了不同的气动配置。事实上，结构的差异取决于满足纵向稳定性要求的方法。基本配置如下。

1. 串联配置

串联配置通过调整设计俯仰角度和前后翼元件的几何形状来解决稳定性问题。这种方法允许以适当的方式改变空气动力学中心的稳定性，同时使用具有最大容量的翼型来利用地面效应。第一个使用串联配置的模型是 1960 年推出的 3 吨级 SM-1 飞行器。虽然一定范围的高度-俯仰参数稳定，但该模型具有较高的起飞速度和飞行刚性。此外，运动稳定性的高度范围变得过于狭窄。串联方案已被 Jörg(德国)成功使用，他开发这种配置多年，并在世界上建造了大部分串联方案飞行器(串联机翼船)。

串联配置的优点是：结构简单，配置调整简单，以确保给定的静态稳定裕度、有效的单通道(节流)控制、小跨度，即与船舶更相似的长梁比。该配置的主要缺点是它只能在静态稳定裕度对俯仰角和离地间隙的组合非常敏感的地面效应中应用。对于小型飞行器，其最大操作高度很小，适航性也受到限制。

2. 翼尾配置

翼尾配置的主要特征是有一个靠近地面移动的大型主翼，以及一个安装在垂直稳定器上不受地面效应影响的水平尾翼，从而使俯仰中心向下游移动。该类飞机计划源于苏联的飞机研发和建设工作，创造了第一代大型 ekranoplan 地效飞行器，其代表是“KM”、“Orlyonok”、“Loon”和“Strizh”。

这种配置的主要优点是：大范围的高度和高度-俯仰组合参数，这样飞行器可以保持稳定的飞行，具有“跳跃”的能力，可以提供高效的转弯动作，并具有在降落时动力增强的能力。大的机翼载荷产生高速(这抵消了由于有效载荷低而导致的运输生产力损失)。这种配置的缺点是：对于高位安装的巨大的尾翼单元(高达主翼面积的 50%)而言，重量非常大，这对飞行器的提升能力贡献很小，同时增加了额外的黏性阻力。与隔离主翼的高升阻比相比，由于非升力面积较大，因此提升阻力比较低(就经济效率而言)。结构重量大导致此空机重量较大。

3. 飞翼配置

飞翼配置的特点是非升力部件显著减少，并且水平尾翼非常小(或不存在)。我们可以看出飞翼布局的特点是利用整个飞行器来产生飞行所需的升力，通过机翼下方的特殊轮廓设计和自动稳定/阻尼系统的应用来提高纵向稳定性。

这种类型的“尾巴”配置是由 Alexeev 在 20 世纪 70 年代提出的。然而，当时他的想法很难实现，然后该计划被放弃了。飞翼配置的浮空飞行器有：Amphistar-Aquaglide(苏联)和在日本测试的 WISE 飞行器。由于机翼部分的智能轮廓，这两种浮空飞行器都具有自然稳定性。在形式上，一些其他飞行器可以用到这种配置(如日本 KAG-3)，即使它们没有“飞翼”型稳定性特征。

这种配置的优点是：有效利用飞行器，最大限度地利用地面效应；低空隙率组分，特别是对于小纵横比的飞行器。缺点是：低范围的高度-俯仰组合参数难以实现纵向稳定性(不使用自动控制系统的情况下)，相对低的操作飞行高度，很难保证水基全翼飞行器的结构完整性；襟翼的低效使用或使用不当可能会降低运动的静态稳定性。

4．复合翼配置

复合翼配置旨在结合飞机结构和飞翼配置的优势，从而在功率增强时实现高起飞效率。“普通”复合材料机翼具有小纵横比的中央机翼，具有高纵横比的端板和侧翼。它采用了剖析主翼下侧的方式来减少尾翼。复合翼的整体纵横比超过了第一代飞行器的主翼，并且远远大于串联配置的纵横比以及现有的“飞翼”(小于 1.5)。这种特性导致升阻比更高，并且与翼部的 S 形成相结合，提供更高的效率和范围。Centroplan 飞行器的小纵横比使功率增强和起飞效率达到了最大化。基于复合翼配置的飞行器属于 MPE 飞行器系列，例如，400t 重并可以搭载 450 名乘客的 ekranoplan MPE 飞行器，其主翼面积减小了约 27%，而航程增加了 3000km。

7.1.3　地效飞行器简史

最早使用地面效应的是莱特兄弟。飞行员在所谓的“缓冲效应”或“煎饼”着陆的掩盖下遇到了地面效应现象。水上飞机 Dornier DO-X 的跨大西洋服务使其有效载荷和航程增加(1930—1931)。常规军用飞机(F105D、B-58、Avro Vulkan)的乘坐和操纵性能得到改善，即使在地面超过五个跨度的距离也经常出现。

第一个设计地效飞行器的应该是 Kaario(芬兰，1935 年)。他的“8 号航空母舰”的特点是有一个小纵横比的机翼，斜靠在板(艉鳍)上和旋转的机翼上，能将空气螺旋桨射流引导到主翼下方。为了提供额外的静态稳定度，Kaario 增加了两个带有稳定面的纵向后梁，如图 7-2 所示。

图 7-2　Kaario 的 8 号航空母舰

动力增强系统的前身是华纳“压缩机”飞机(美国，1928 年)，如图 7-3 所示。该设计采用鸭式布局，包括两个强大的风扇，使空气在飞行器的圆顶状底部下方。华纳“压缩机”飞机是第一架使用独立起飞和巡航引擎的飞机。

图 7-3 华纳“压缩机”飞机

公羊翼概念由 Troeng(瑞典，20 世纪 30 年代)指出，如图 7-4 所示。Troeng 的矩形翼飞行器的特点是：①借助特殊浮子增强静态稳定性；②使用螺杆螺旋桨；③在机翼后缘使用小型水翼来确保设计巡航模式下的纵向稳定性。

图 7-4 Troeng 的公羊翼

洛克希德自 1960 年以来一直参与地效飞行器的研发。1963 年，他设计了一艘装有端板的小型双座船(Koryagin)。它由两个弓形水力传导，以获得更好的纵向稳定性。洛克希德还以设计了具有地面效应的大型飞行双体船而闻名。该飞行器由襟翼、副翼和尾翼单元包括方向舵与升降舵来实现稳定和飞行控制，船体和机翼运输货物。后来，洛克希德-格鲁吉亚研究了一艘 13.62 亿磅的机翼，该机翼被设计成一种物流运输，能够在海况 3 级环境下，载重约 200 吨，以 0.40 马赫的巡航速度在 4000 海里以上的开阔海域进行运输。

7.2 地效飞行器项目

7.2.1 中国研究项目

1967 年，中国地效飞行器的研发和设计在中国船舶科学研究中心(CSSRC)启动。从那时起，在超过 30 年的时间里，共有 9 辆小型载人试验飞行器在湖泊和沿海地区进行了设计和试验。XTW 系列采用翼尾布局，主翼具有前向扫掠，如 Lippisch 的设计，如图 7-5 所示。

图 7-5　XTW-1 系列飞行器

1996 年，CSSRC 研发了 XTW-2，XTW-3 和 XTW-4 地效飞行器。该系列的典型代表是 XTW-4，它将 XTW-2 稍微改进，以符合海上试验的特定要求。这艘运载 20 人的地效飞行器于 1999 年秋季在长江首次测试。该飞行器包括：主船体(浮子)、由两个小浮子支撑的主翼、两个承载高位尾翼的垂直稳定器。在某种程度上，飞行器的配置属于翼尾式配置。主翼具有向前扫掠功能。两个 P & WC PT6A-15AG 涡轮螺旋桨发动机配有 MT 的 5 叶片可调螺距螺旋桨，安装在主翼的前缘。因此，滑流能有效地用于辅助起飞。

20 世纪 80 年代初，中国船舶及海洋工程设计研究院(MARIC)开始研发两栖地效飞行器(AWIG)的产品。大约测试了 80 个模型，以研究最佳机翼轮廓、空气通道的配置、船艏推进器的位置、尾翼的布置等。在上海郊区的湖上测试了 30kg 的自行式无线电控制模型。由于该测试得到的结果是可接受的，所以 MARIC 开始研制大型船舶 AWIG-750，它的最大牵引质量为 745kg，长度 8.47m，跨度 4.8m，高度 2.43m，如图 7-6 所示。发电系统由内燃发动机组成，其中两个用于升降，两个用于推进飞行器。每台发动机驱动一台额定功率为 30 马力(1hp=745.7W)的管道式推进器 DT-30，速度为 6000r/min。该飞行器能够以 0.5m 的波浪起飞，最高时速为 130km/h。它展现了从水中到岸边往返的预期(两栖)能力。

图 7-6　AWIG-750

1995 年，中国船舶工业集团公司委托研发了“天鹅一号”的 20 座 AWIG-751。该飞行器于 1997 年 6 月完工，其牵引力为 8.1t，长宽高尺寸为 19m×13.4m×5.2m，静水中的最大巡航速度为 130km/h。它有三个航空型活塞发动机，其中两个用于标准升降机的 257kW HS6E 发动机，一个用于推进的 210kW 的 HS6A 发动机。标准发动机驱动了两个弓形导管四叶片空气螺旋桨，巡航发动机驱动了一个双叶片可变螺距螺旋桨。与 AWIG-750 相比，它有几个新特征：增加了主翼的跨度，采用复合机翼，将叶片与襟翼组合使用以增强纵向稳定性，采用 CHIBA 复合材料可减少结构重量。

经测试证明，总体符合设计要求，但也显现出一些缺点，即弓形螺旋桨的轴驱动器太长、有效载荷较低、离地间隙低于预期。因此后续飞行器机型 AWIG-751G(天鹅二号)增加了尺寸，改进了标准发动机布局和复合机翼。

7.2.2 德国研究项目

Rein-Flugzeugbau 公司的前技术总监 Hanno Fischer 成立了 Fischer Flugmechanik 公司，并扩展了 Lippisch 飞行器的设计理念，开发制造了一款名为 Airfish FF1 / FF2 的双座飞行器。与 X112 和 X114 不同的是，Airfish 针对地面效应而设计，由玻璃钢材料制造，在 1988 年的测试中，速度为 100km/h，仅为发动机功率的一半。1990 年，Fischer Flugmechanik 测试了一款 4 座飞行器 Airfish-3，它比 Airfish FF2 重 2.5 倍，飞行速度为 120km/h，能够覆盖 370km 的范围，如图 7-7 所示。这个飞行器的长度为 9.45m，宽度为 7.93m，操作间隙为 0.1～1m。

图 7-7　Airfish-3 飞行器

基于 Fischer Flugmechanik 开发的 Airfish 系列的设计重新出现在 Flightship 8(最初被命名为 Airfish 8 的 FS-8)中，如图 7-8 所示。FS-8 由德国的 Airfoil Development 有限公司开发，并于 2000 年 2 月在荷兰首次飞行。该飞行器的牵引力为 2325kg，长度为 17.22m，宽度为 15.50m，高度为 4m，Flightship 8 飞行器可以载 8 人，其中包括两名机组人员。起飞时海面波浪的最大高度限制在 0.5m，但在巡航时它可以通过 2m 的波浪。FS-8 由玻璃钢材料制成，额定功率为 330kW，巡航速度约为 160km/h，续航里程为 365km。

图 7-8　Flightship 8 飞行器

体型较大的 Flightship-40(FS-40)飞行器称为 Dragon-Clipper，其通勤版本可容纳 40 名

乘客，相当于具有5t的承载能力。较大的飞行器长度为30m，25m的翼展可以通过折叠小翼减小到20m，便于岸上操作。主要建造材料是铝，功率1000kW的Pratt和Whitney涡轮螺旋桨柴油发动机将使巡航速度提高到约225km/h。最大起飞浪高1.2m，翼展跨度增加，可在水面4m以上的海域飞行进行水上作业。Fischer Flugmechanik和Airfoil Development公司的FS-8系列的创始人宣布了一项结合地面效应和静态气垫技术的新设计HW-20飞行器的提议。HW-20(Hoverwing)的设计采用简单的伸缩灵活系统，用来在主船体配置的双体船舷舱之间保留气垫。这种静态气垫仅在起飞时使用，在无缝过渡到真正的地面效应模式之前，飞行器能够以最小的功率加速，如图7-9所示。

图7-9　Hoverwing-20飞行器

7.2.3　俄罗斯研究项目

复合机翼意味着将飞机起重区域的功能细分为两部分：一部分(中央)用来实现动力增强模式，另一部分(侧翼)用来提高巡航中的效率和纵向稳定性。在这种情况下，影响稳定性的有三种因素，分别为主翼中心部分的特殊剖面、水平尾翼(尽管相对较小)、侧翼的几何形状和位置。MPE系列(设计者General D. Synitsin)的设计充分利用了以上特性，其牵引力为100~400t，如图7-10所示。

图7-10　MPE-400飞行器

Ekranoplan Amphistar由“技术与运输”公司(主任兼首席设计师D. Synitsin)于1995年开发和制造。1997年，该飞行器被授予俄罗斯联邦航运登记证书，是一种动态气垫的小飞行器。其最大牵引质量为2720kg，长宽高尺寸为10.44m×5.9m×3.35m，巡航速度为150km/h，行程可达450km，适航度约为0.5m，巡航速度的转弯半径大约为65个船体长度。在水中，转弯半径大约是船体的长度。该飞行器的改进版本为名叫Aquaglide的飞行器，如图7-11所示。Synitsin开发了一系列Amphistar-Aquaglide型飞行器，如图7-12所示。

图 7-11　Aquaglide-5 飞行器

Figure 7-12　Aquaglide-50 飞行器

7.2.4　美国研究项目

在 20 世纪 90 年代初期，美国 AEROCON 公司开发了 Aerocon Dash 1.6 项目，如图 7-13 所示。这种庞大的飞行器具有以下物理特性：牵引力为 5000t，有效载荷数为 0.3588，机翼载荷为 258lb/ft^2，巡航速度为 400 节，巡航高度为 12ft。

过去一段时间里，洛克希德·马丁航空系统公司(LMAS)对海基飞机的发展进行了研究。LMAS 呼吁研发混合动力飞机，并寻找合适的混合动力解决方案，从而产生了一系列设计方案，包括水上飞机、浮子飞机和类似地面效应的复合表面效应飞机(SEA)，如图 7-14 所示。SEA 结合了多种表面效应技术，将被用于基于海洋的混合飞机设计——地效飞行器、水上飞机和水上飞机船体。据 LMAS 称，这种概念在目前的飞机技术中是可行的，可以提供高达 400 节的速度和 400t 的承重能力。

图 7-13　Aerocon Dash 1.6 飞行器

图 7-14　SEA(地面效应飞行器)

根据波音前沿公司报道(在线，2002 年 9 月，第 01 卷，第 05 期)，当时波音幻影工程公司开发了一种名为 Pelican(鹈鹕)的高容量货机的飞行器设计概念，如图 7-15 所示。它有一个大纵横比的机翼，翼展 500ft，翼面积超过 1 英亩，是当时世界最大型飞机 An-225 的两倍，它可以运输多达 1400t 的货物。

Pelican 具有较长的跨海洋航程，可以飞行至海拔 20ft(基于跨度的相对离地间隙大约 20/500 = 0.04 的高度)，也可以飞行在 20000ft 或更高的高度。适用于需要速度的全球范围和高吞吐量的商业与军事运营方面。波音高级升降机和油轮的战略发展高级经理 John Skoupa 指出，Pelican 是唯一确定的美国陆军能够在 5 天内部署一个师或在世界任何地方 30 天内部署 5 个师实现部署转型目标的手段。它可以在一次出击中携带 17 辆 M-1 主战斗坦克。

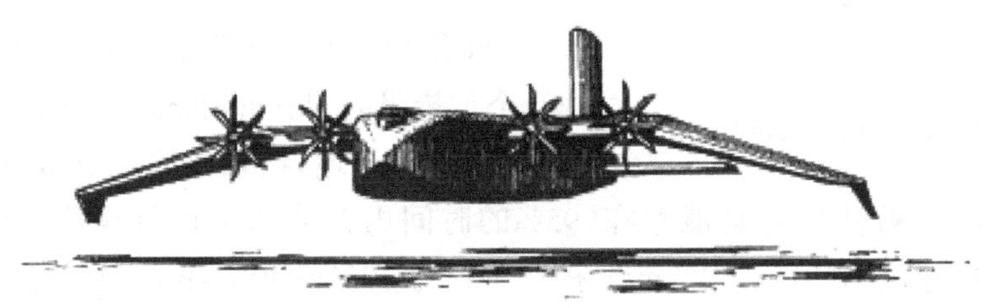

图 7-15　Pelican 飞行器

7.3　地效飞行器的应用

7.3.1　民事用途

根据 Belarin、Volkov 和 Hooker 报告的初步分析，地效飞行器具有发展商业计划的前景，用于运载乘客或货物，也用于旅游、休闲以及特殊需求，如搜索行动。

1．搜救行动

1989 年 4 月 7 日在挪威海的“共青团员”核潜艇和 2000 年 8 月 12 日在巴伦支海的核潜艇“库尔斯克”所发生的悲剧至今记忆犹新。对现有水上救援方式的分析表明，水面舰艇无法迅速地到达灾难现场，而飞机无法进行有效的救援行动，因为飞机无法在沉船附近着陆。与 ekranoplan 相比，大多数现代水上飞机也具有较低的承载能力和适航性。地效搜索救援飞行器 Spasatel 由下诺夫哥罗德的 Volga 工厂建造。

Spasatel 飞行器如图 7-16 所示，基于 Loon 型 ekranoplan，结合了所有已知的海上救援方式(搜救飞机、直升机、飞艇)的特点。其巡航速度预计在地面效应的 400～550km/h 范围内，而地面效应的巡航速度可达 750km/h。远离下方表面飞行时的海拔可达 7500m，搜索模式下高度约为 500m。该飞行器可以在高达 3.5m 的波浪中着陆并进行救援行动。还可以在波涛汹涌的海面上航行，波浪高度达到 4m。Spasatel 可在 3000km 范围内运行，自主运行时间高达 5 天，最多可容纳 500 人。在决定开发 Spasatel 之前，已经对可用的导弹载体“Loon”进行了几次实验，以评估 ekranoplan 作为救援飞行器的能力。这些实验表明，ekranoplan 具有的特性可用于水上救援行动。特别是当漂浮在水面时，飞行器前部自然迎风移动。由于飞行器的主翼部分(尾部)浸没在水中，因此在其后面形成相对平静的水域。

图 7-16　Spasatel 搜救地效飞行器

2．全球海上救援系统

在公海上制定有效措施是世界各国普遍关注的问题。经验表明，在海上发生沉船和生态灾害时提供援助是非常困难的。由于不利的气象条件，水上飞机的使用通常受到限制，而直升机的使用仅限于沿海地区。以前，海上救援(打捞)的主要手段是船舶，而船舶要靠近灾害区域存在偶然性，几乎不适合救援。

因此全球海上救援系统被提出，该系统包括 50 个重量级的 ekranoplans，以世界各地 12 个重点基地港口为基地。该系统的每个 ekranoplan 都具有较高的起飞/着陆适航性，并且能够在 95%的时间内在公海上进行操作。该系统的每个 ekranoplan 的巡航速度为 400～500km/h，操作半径为 3000～4000km。当海运速度为 15 节时，飞行器可在海面上长时间航行。救援飞行器应该为灾区提供多种救援手段，包括木筏和自行切割工具，可能还有直升机和深海潜水器。

3．水平发射航空飞机

根据俄罗斯圣彼得堡国立航空航天仪器仪表大学和日本东京的武藏工业大学联合开发的项目，无人驾驶的自行推进的 ekranoplan 地效飞行器应该可以载重，可以接近声速的一半，并将 600t 火箭飞机发射到低地球轨道(水平发射)，如图 7-17 所示。将更多的有效载荷发射到近地轨道并拓展航空航天运输系统的功能，这是 21 世纪新空间项目开展的主要任务之一。

图 7-17　水平发射飞行器

7.3.2　军事用途

对已知项目和未来海军应用的分析已经证实，以上列出的地效飞行器属性以及速度快、雷达能见度低、海上保持能力强、可与类似大小船舶相媲美的有效载荷等高突发因素，冲刺速度特征和在开阔海洋中的漂浮能力，使地效飞行器成为完美的多任务武器平台，可以提前部署并通过招标进行操作。

海军 ekranoplan 可用作对抗陆地和海上目标的攻击战武器、战术和战略巡航导弹的发射平台、航空母舰和两栖攻击运输飞行器。在中等海况下容易降落，使利用 ekranoplan 作为能够有效部署水听器或拖曳阵列的反潜战机成为可能。地效飞行器还可用于各种侦察和运输任务。地效飞行器也符合世界各地锚地作业理念，以保持处于前沿状态。

1．两栖战争

地效飞行器的速度、承载能力和低空巡航能力使毁灭性的突袭成为可能。分析人士还注意到，与冲突期间地面部队到达的天数相比，地效飞行器可以在几个小时内从英国抵达马尔维纳斯群岛。标准地效飞行器两栖作战的主要困难是人员和设备的着陆。由于减少结构重量是实现有效地效飞行器飞行的关键因素，因此飞行器在不降低其巡航性能的情况下不能进行海上行动。

两栖攻击艇的一个例子是俄罗斯名叫“Orlyonok”的 ekranoplan。尽管“里海怪物”这个绰号很吓人，但它并不是战机，而是一个巨大的飞行试验台，Orlyonok 是第一个专为军事目的设计的 ekranoplan。载重 20t 的飞行器货舱长 24m，宽 3.5m，高 3.2m。为了实现大型货物和重型军用车辆(如坦克和装甲运输车)的登船和离船，Orlyonok 具有独特的摆动式船艏设计。

2．海上升降机

ekranoplan 在海上运输方面作用较大。正如我们所想的那样，为了能在高海况下飞行，跨洋地效飞行器的体积要非常大，至少净重 900t。即使如此，据估计，这样一个地效飞行器可以承载比三架 300t C-5 飞机的货物还要多的重量，而燃料仅使用 60%。地效飞行器兼具了传统空气升降机和慢速水面运输的特点。与飞机不同的是，地效飞行器不依赖海外基地。同时与船舶不同的是，地效飞行器海上起重机速度快，不需要护航，对鱼雷和地雷的抵御能力较强。

3．侦察和巡逻

大型地效飞行器应用最少的是侦察或巡逻。由低空作业产生的极限范围将大大减小截取雷达或信号的范围，从而大大减小区域覆盖范围，导致可能无法有效地利用该平台。即使在对船舶的打击战中，地效飞行器也可以从其他平台获取目标信息。

参 考 文 献

昂海松，童明波，余雄庆，2008. 航空航天概论. 北京：科学出版社.

库利，吉勒特，2008. 飞艇技术. 王生，等译. 北京：科学出版社.

《世界飞机手册》编写组，2012. 世界飞机手册. 北京：航空工业出版社.

BOSSERT D E, MORRIS S L, HALLGREN W F, et al., 2003. Introduction to aircraft flight mechanics. The American Institute of Aeronautics and Astronautics, Inc.

BRANDT S A, BERTIN J J, STILES R J, et al., 2004. Introduction to aeronautics. The American Institute of Aeronautics and Astronautics, Inc.

ILLMAN P E, 1992. The pilot's handbook of aeronautical knowledge. New York: McGraw-Hill.

LIN L, LGOR P, 2009. A review of airship structural research and development. Progress in Aerospace Sciences, 45(4-5):83-96.

ROZHDESTVENSKY K V, 2006. Wing-in-ground effect vehicles. Progress in Aerospace Sciences, 42(3):211-283.